OLD DESERET LIVE STOCK COMPANY

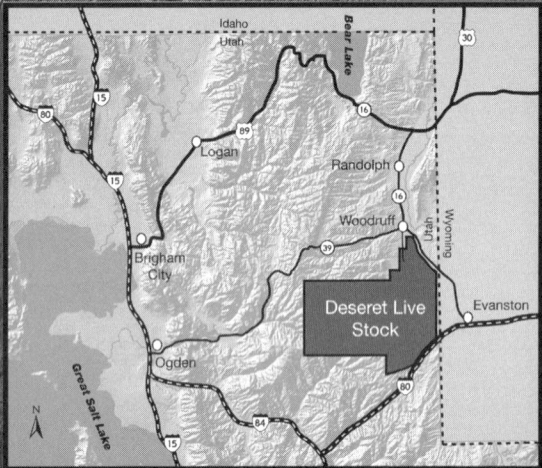

Eastern
Nevada Range

Deseret
Live Stock

Skull Valley
Winter Range

Idaho
Utah

Bear Lake

30

80

15

89

Logan

16

Randolph

15

16

Woodruff

39

Utah

Wyoming

Brigham
City

Deseret Live
Stock

Evanston

Ogden

80

N

Great Salt Lake

84

15

OLD DESERET LIVE STOCK COMPANY
A Stockman's Memoir

W. Dean Frischknecht

Utah State University Press
Logan, Utah

Utah State University Press
Logan, Utah 84322-7200
www.usu.edu/usupress

Manufactured in the United States of America
Printed on recycled, acid-free paper

ISBN: 978-0-87421-711-7 (cloth)
ISBN: 978-0-87421-712-4 (e-book)

Library of Congress Cataloging-in-Publication Data

Frischknecht, W. Dean.
  Old Deseret Live Stock Company : a stockman's memoir / W. Dean Frischknecht.
    p. cm.
  Includes index.
  ISBN 978-0-87421-711-7 (cloth : alk. paper) -- ISBN 978-0-87421-712-4 (e-book)
  1. Frischknecht, W. Dean. 2. Sheep ranchers--Wasatch Range (Utah and Idaho)--Biography.
3. Ranch life--Wasatch Range (Utah and Idaho) 4. Deseret Live Stock Company--History. 5.
Wasatch Range (Utah and Idaho)--Biography. I. Title.
  SF375.32.F74A3 2008
  636.3'01097922--dc22
  [B]
                                    2008007384

Contents

PROLOGUE

THE DESERET LIVE STOCK COMPANY in northern Utah was organized and incorporated in January 1891. "Deseret" is a Book of Mormon word for "honeybee," a symbol of industry. Brigham Young and many of the early Utah pioneers hoped that the Territory of Utah would become a vast state named Deseret. Those early settlers knew that if they were to succeed in the high dry country of the Great Basin, they would have to work and cooperate like honeybees. In 1896 the Territory of Utah became a state named Utah, in recognition of the ancestral home of the Ute Indians. The Deseret Live Stock Company, though, retained the name early Mormon Utahns had applied hopefully to a large section of the West.

Before the company was incorporated, neighboring ranchers from Davis County and other parts of northern Utah laid the groundwork for a large cooperative enterprise. Led by members of the Hatch and Moss families, a number of woolgrowers sought to join together in a large outfit with a hired manager. After they had determined the outside limits of a desirably sized ranch in the northern Wasatch Mountains, property owners within the tentative, desired area would be invited to join the larger entity. This idea caught on and, at the time of incorporating, successfully attracted ninety-five stockholders, who jointly owned twenty-eight thousand sheep, a number of horses, and some sheep camp equipment, valued at $90,000. The Deseret Live Stock Company was incorporated with seventy-five thousand shares. The incorporation papers stated that the purpose of the company was to "promote stock raising, slaughtering, butchering, mercantile business, and other commercial enterprises in the Territory of Utah and other states of the U.S."

The Deseret Live Stock Company was prosperous, and after six months of successful operations, a 10 percent dividend was paid to all stockholders. The Moss and Hatch families were the largest stockholders, and held the top managerial and executive positions in the company. Substantial profits allowed the company to increase the number of sheep it ran to fifty thousand, to get into the cattle business, and to purchase additional land for grazing and the production of farm crops such as hay and grain. Much of the land was purchased in northeastern Utah, from the checkerboard grants the Union Pacific

1

Railroad had received from the federal government to subsidize construction of the transcontinental railroad. The company acquired other land from the state of Utah and from private individuals. Following the entry of the United States into World War I, livestock producers such as Deseret profited greatly from the demand for wool, beef, and mutton by the U.S. government.

In 1917 the Deseret Live Stock Company purchased the Iosepa Ranch in Skull Valley, Tooele County, Utah, from the Church of Jesus Christ of Latter-day Saints (LDS, or Mormon, Church). This fertile, irrigated ranch would serve as winter headquarters for grazing sheep and cattle seasonally driven out of the high mountains and for the production of hay and grain. This historic ranch in Skull Valley had been purchased by the LDS Church from John Rich in 1890 for $58,000. Iosepa had been created in 1889 as a colony for a group of Hawaiian converts to the church who wished to move to Utah. Church officials looked over several possible locations not too far from Salt Lake City before settling on the Skull Valley Ranch. The LDS Church helped build the town of Iosepa there, which became home to 230 Hawaiians. The town was surveyed and laid out in square blocks, and a pressurized water system was installed. A store, schoolhouse, church, and many homes were built. However, the church built a temple in Hawaii during the years 1916 to 1919 on Oahu, across the island from Honolulu. Most of the Hawaiians at Iosepa chose to move back there, although some of the younger colonists moved to Salt Lake City. The church had owned the property at Iosepa and paid the Hawaiians a living wage, which enabled the community to last for twenty-eight years.

Soon thereafter, the Deseret Live Stock Company bought the Tom Jones ranch, which adjoined company property along the Utah/Wyoming line, for something over $200,000. This was followed by the 1918 purchase of the neighboring Neponset Land and Livestock Company in northeastern Utah for another $200,000. Deseret continued to prosper and buy more land; by the 1920s it owned nearly 225,000 acres and was said to be the largest private landowner in Utah.

After the stock market crash inaugurated the Great Depression in autumn 1929, the early 1930s value of livestock dropped to less than 30 percent of its 1929 value. The days of high profits and dividends had vanished. Sales of livestock and wool would not pay production expenses. The company was unable to pay dividends from 1930 through 1938. During this time, several stockholders wanted to sell their stock. Most of this was purchased by an original stockholder and member of the board of directors, James H. Moyle, and his family. Moyle eventually owned two-sevenths of the company, nearly one-third. He had served as undersecretary of the U.S. Treasury in

Walter Dansie, general manager, left, and James H. Moyle, president, of the
Deseret Live Stock Company (circa 1940).

President Woodrow Wilson's administration. His son Walter Moyle, a lawyer
in Washington, D.C., became owner of one-seventh of Deseret.

In 1933 the Deseret Live Stock Company borrowed $370,000 from the
federal Regional Agricultural Credit Corporation. The company's board
of directors agreed that the lender should appoint a new general manager,
Walter Dansie, who was an experienced livestock man and broker. He replaced
William Moss, who had been the manager for over forty years, but was in his
declining years and not in good health.

Dansie introduced several worthwhile changes. He closed the three retail
stores the company operated. Most of the mercantile business was through a
big store in Woods Cross, Utah; however, it had been customary to let customers charge their purchases, and the store consequently was losing money.
Steps were taken to collect overdue accounts, and the store was closed. Dansie
moved the Deseret Live Stock business office from Woods Cross to Salt Lake
City, and the company concentrated on livestock production. Sheep camps
and ranches would thereafter order food using "grub order" sheets, which
listed several dozen food items that were available. Each camp and ranch
ordered enough food to last six weeks. ZCMI Wholesale Grocery in Salt Lake
City filled the orders and delivered them to central locations out on the range
on a designated "grub day."

Will Sorensen and his wife, Margaret, rode horses from the summer
range to the Home Ranch, eighteen miles each way, to pick up fresh
garden vegetables (circa 1940).

Eventually, James H. Moyle became president of the Deseret Live Stock
Company. He held that office when I became acquainted with the company, at
shearing time in June 1943, while I was working on my master's degree.

Kathryn and I were married in February 1941, almost a year before Pearl
Harbor was bombed by the Japanese on December 7, 1941. Because I was a
married man, I received a "3A" classification in the U.S. military draft. This
gave me time to graduate from (now) Utah State University with a B.S. degree
in 1942, and to follow that with graduate work at USU. That summer I worked

in the wool laboratory on campus and became an experienced wool grader. At that time Charlie Redd, a prominent rancher from La Sal, in southeastern Utah, was chairman of the Board of Regents for USU. He made arrangements for me to be at his ranch during spring shearing in 1943, and to grade the ranch wool clip of twenty-five thousand fleeces. This was twenty thousand fleeces from his sheep and five thousand fleeces belonging to his cousin Joe Redd, who worked with him. It was a great experience for me, and he offered me permanent employment. He could tell that I had grown up working with sheep, cattle, and horses on our family ranch; however, I could not accept employment at that time.

When I got back up to Logan, home of USU, where Kathryn and I were living, Dr. Fred F. McKenzie, head of the Department of Animal Science, and Professor Alma Esplin, sheep extension specialist, had made arrangements for me to be at the Deseret Live Stock Company during shearing, which started June 12, 1943. My job there was to examine each of the company's forty thousand ewes just before they entered the shearing shed, and classify the top twenty thousand that would be bred to whitefaced rams to raise replacement-ewe lambs. The twenty thousand lower-quality ewes were to be bred to black-faced Suffolk or Hampshire rams to produce the heavier muscled, meatier market lambs. Sheep foreman Will Sorensen had constructed two alleys, side by side, fifty yards long and three feet wide, which would be filled with sheep for me to examine. The two alleys were arranged so I always had sheep to work on. The lower-quality ewes received a mark on the bridge of the nose from yellow grease chalk I carried. I had to work fast. As I finished an alley full of sheep, they were driven into the shearing shed where there were twenty-two shearers in a long line, working with power-driven shearing equipment. After being shorn, those ewes with a yellow mark on the nose were given a special small, dark yellow paint brand back of the shoulder, so they could be identified during the breeding season in December.

Walter Dansie offered me permanent employment at that time, but I could not accept because I had to finish my M.S. degree and thought I might have a military obligation coming up. I was honorably released from the U.S. Marine Corps on May 31, 1946, having been a replacement for an outfit that had fought on Iwo Jima and having served during the occupation of Japan.

When I returned to the Deseret Live Stock Company as a permanent employee, just in time for shearing to start in June 1946, President James H. Moyle had died. He was replaced in that position by his son Henry D. Moyle, a lawyer and businessman in Salt Lake City. As I started work, I was informed that the Deseret Live Stock Company produced the largest clip of wool in the world from an individual ranch.

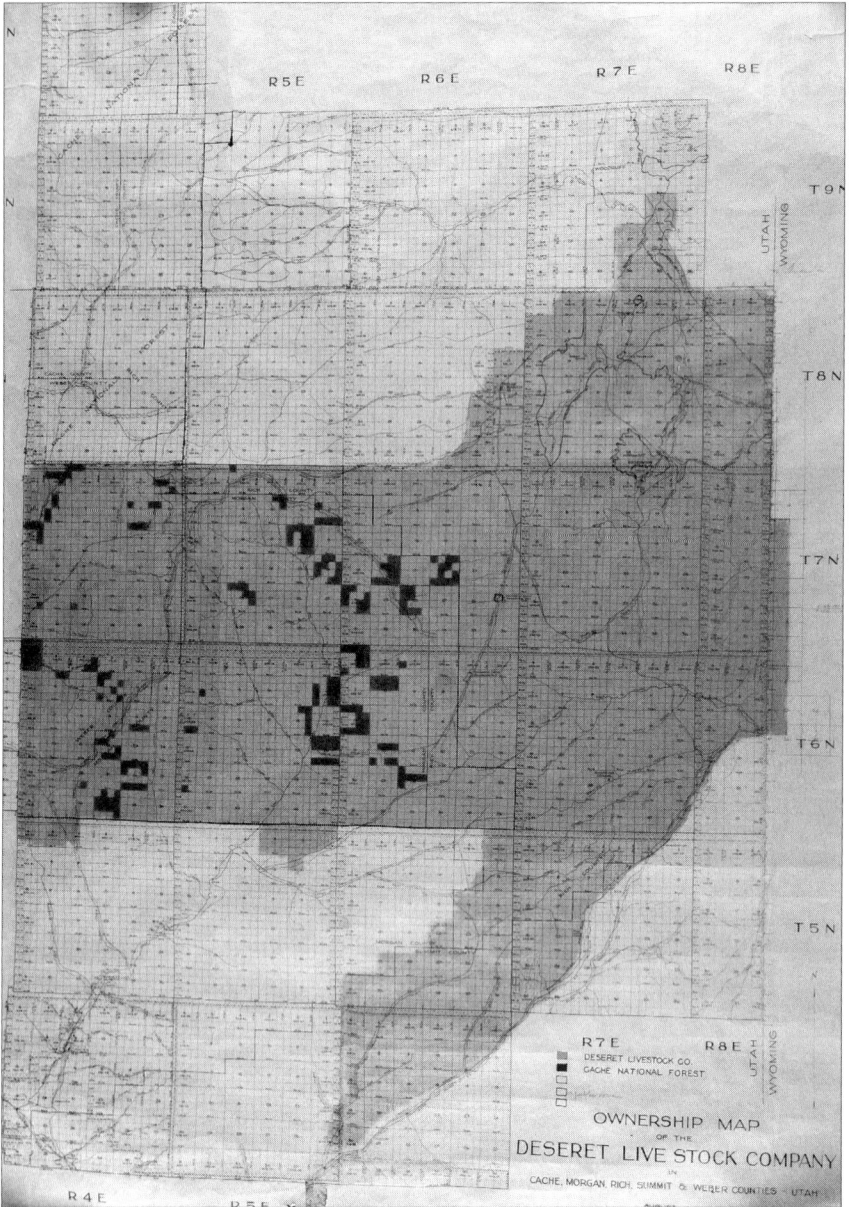

Home Ranch

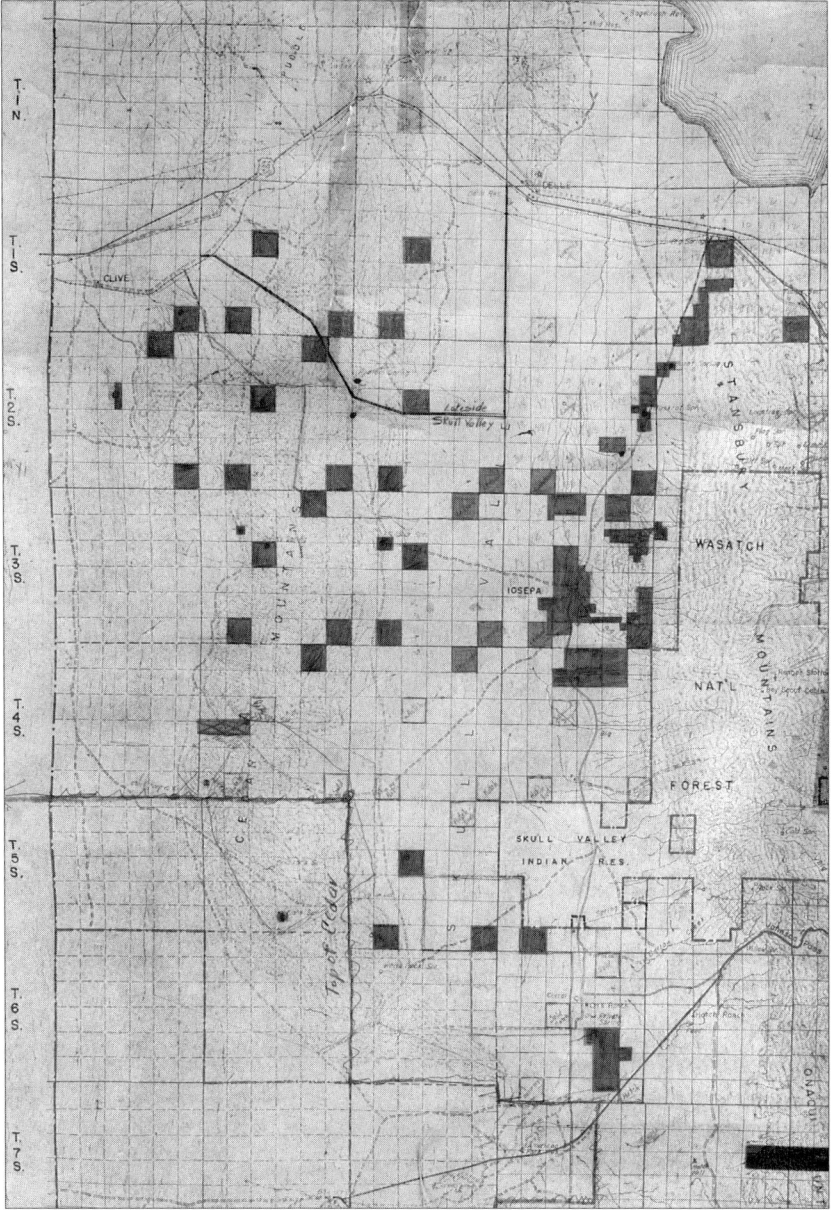

Skull Valley Ranch

Walter Dansie on the porch of our home at the shearing corral.

EXPANDING HORIZONS

THE ELDERLY WILL SORENSEN WAS at my side, and I was at the wheel of the black Chevrolet pickup as we wound our way up the dirt road. I was listening, asking a question now and then, and also mulling over what Walter Dansie, general manager of the Deseret Live Stock Company, had told me a few days earlier in the company office in Salt Lake City. My major job would be to take over the day-to-day management of their forty thousand sheep, plus their lambs, in twenty-six summer herds on the 250,000-acre Home Ranch in northeastern Utah. The ranch was 90 percent privately owned, and 10 percent federal lands. About five thousand cattle and a few dozen yearling and two-year-old horses grazed during the summer on some of the range that they shared with the sheep, so my responsibility included being vigilant about the condition of our ranges and the welfare of our cattle and some of our four hundred horses.

We drove over the crest of a grass-covered hill and saw two cowboys, a hundred yards downhill, rope the head and hind feet of a yearling Hereford heifer and stretch her out on the grass. I let the pickup roll to an easy stop just a few yards from the heifer—a heifer that early that morning had been too inquisitive about a porcupine and now had a snoot full of quills. The horses kept the ropes taut as the cowboys quickly dismounted and went to work. One held the heifer's head to the ground as the other loosened the rope and used a pair of pliers to remove the quills from her muzzle. She didn't struggle; she knew she was being helped. Soon she was released, got to her feet, and shook her head from side to side as she trotted back to her bunch. The cowboys remounted and were coiling up their ropes when one called to Will and thanked him for getting word to them about this heifer. Will told them he was pleased they took care of her that day; if left until the next day, some of

Author's Collection
Will Sorensen's wife, Margaret, who died in 1945,
shown here at the shearing corral.

the quills would have broken off so they couldn't be pulled out, and she would have been in big trouble. Will told me that a porcupine wouldn't slap another animal with its tail unless it felt threatened, and it was usually some young and foolish critter that got too close and got itself hit. I replied that while growing up in Manti, I had seen two town dogs and two yearling cougars with porcupine quills stuck in their faces, but never any sheep or cattle. He glanced at me, and I let the pickup proceed down the gently sloping, wide-bottomed Heiner's Canyon, part of the Home Ranch.

Hereford heifers, Heiner's Canyon, 1946.

"Will" was sixty-three-year-old W.H. "Will" Sorensen, highly respected longtime sheep foreman for Deseret. His six-foot frame still handled his two hundred pounds, but in a slightly stiff-at-the-knees, mature manner. His large gray Stetson and eyeglasses added to his stature and bearing. He told me the success of the "Live Stock Company" had been his whole career, but that it was different out here now. Will said that when you lost your mate, the days and nights got long and lonesome. His wife, Margaret, had died a year earlier; they had no children, and now he had asked to be replaced because life was too lonely on this vast range.

I, twenty-six-year-old Dean, was the man who had been hired to replace him. I was the same height as Will, but broader of shoulder and thicker bodied. My attractive auburn-haired wife, Kathryn, a year younger, and our two children, "Bill" Dean, age three, and Diane, two, arrived with me earlier that day, June 13, 1946, at the shearing corral, the spring sheep headquarters of the Deseret Live Stock Company, three miles northwest of Wahsatch, Utah, near the Wyoming state line.

Will informed me that we had six hundred yearling Hereford heifers summering in Heiner's Canyon. He went on to explain that this was a 26,000-acre fenced pasture on the southeast perimeter of the Home Ranch. Company policy was to keep the yearling heifers away from the bulls—not let them get bred until they were two-year-olds, so they would have their first calves at three. The next summer these heifers could run with the bulls and cows on the main part of the ranch. We usually went into the winter with 5,500 cattle. That included 4,600 over at the Home Ranch headquarters near Woodruff, and 900 down in Skull Valley south of the Great Salt Lake, our winter headquarters for the sheep. The winter sheep range was also 250,000 acres, but instead was 10 percent private, and 90 percent federal.

As we drove on down Heiner's through scattered groves of quaking aspens, I was surprised to see two reservoirs full of water located only two hundred yards apart. Each reservoir was about one hundred yards in diameter. I stopped the truck and asked Will why they had built these two dams so close together. Will started to chuckle, then explained that he wanted to build the dam where the upper one now was. The engineer was one of those college-educated smart guys, and he wanted to build it where the lower one was, so that's where they had to build it. Then the beavers came and built the upper one exactly where Will wanted it. He complained that those educated college people lacked practical experience.

He was right about some of them, but some did have experience. I just smiled, since I was college educated, with B.S. and M.S. degrees in Animal Science from Utah State Agricultural College (now Utah State University) in Logan, Utah, but was also ranch-raised, tending a large herd of sheep, a small herd of cattle, and a dozen excellent horses. Will went on that he was a big kid and had to drop out of school and go to work before finishing the fourth grade, but he had a lifetime of hard work and practical experience; he knew what had to be done, and how to do it. Walter Dansie had told me that I would have to be a diplomat to work with Will, but added, "just remember that he and I and our board of directors handpicked you to replace him."

I was no stranger to the Deseret Live Stock Company. I had worked with Will three years earlier, during the three weeks of shearing when, as part of my master's program, I selected the top twenty thousand ewes to be bred to whitefaced rams to produce replacement ewes, and the remaining twenty thousand to be bred to Suffolk and Hampshire rams to produce blackfaced market lambs. I had been offered permanent employment with the company at that time, but didn't accept because I had to finish my M.S. degree, and later might have a military obligation. Early in January 1945, I became a U.S. Marine, and I went through boot camp in San Diego. I had been in the Pacific area for a short time and was part of a replacement group for an outfit that had fought on Iwo Jima. I had served in occupied Japan since early autumn 1945 and was honorably released from the U.S. Marines on May 31, 1946, at Treasure Island near San Francisco. Now, less than two weeks later, I had accepted an offer to cast my young family's future with this great ranching empire. I knew what we were getting into.

Walter Dansie had told me that the Deseret Live Stock Company was created in 1891 in what was then Utah Territory, when a few neighboring ranchers combined their holdings into one large unit and incorporated. Dansie also said that "deseret" was the old Biblical Hebrew word for "honeybee," the symbol of industry. It is actually a Book of Mormon word, and is the name Brigham Young and some of the early pioneers hoped would be the name of their state. In 1896, five years after the company was created, Congress granted statehood but named the new state Utah, as it was the ancestral home of the Ute Indians. By then the company was well known, so it didn't change its name to "Utah Live Stock Company."

Mr. Dansie said he had been manager since 1933 and would retire in nine or ten years at about age sixty-five. The future was bright for the right young man. My major responsibility would be the day-to-day management of over forty thousand sheep, an opening into the company available to few men. I was happy for this opportunity, proud to be there, and had a promising future in a successful Deseret Live Stock Company. It was a heavy responsibility; however, I would be taught by two able men, Walter Dansie and Will Sorensen. It would bring personal growth, and I could become "the right young man." A few years of tough living on the range for our young family would eventually pay off. This was one of the world's great ranches. "General manager of the Deseret Live Stock Company" was a worthy goal attained by few.

Will told me we should go on down to the mouth of Heiner's and get out on the highway. We would go back east through Wahsatch and then over to the shearing corral. We didn't want to be late for supper. He then pulled a cigar out of his vest pocket, cut off a piece not quite two inches long, stuffed it in the bowl of his pipe, and lit up. I said, "Will, you're the only person I've ever seen cut a cigar into three or four pieces and smoke them in a pipe." He told me that three years ago that he had a cancer cut off his lip, and it was caused by years of smoking cigars. Will explained how the Doc told him to quit smoking on account of the lip irritation, but if he wouldn't quit smoking, then a pipe would be best for the lip. He did not like pipe tobacco, so this cigar in the pipe was his own idea. This way he didn't have to smoke a whole cigar at a time and didn't waste any.

Wahsatch, Utah, just eleven miles west of free-wheeling Evanston, Wyoming, boasted a Union Pacific railroad station and was home for the section crew and their families; "population an even dozen, more or less," according to the railroad agent. The U.P. track ran in an almost north-south direction, parallel to U.S. Highway 40 as it passed through Wahsatch. The

Keith Moss McMurrin Collection

The shearing corral and sheds, where fifteen hundred sheep could be housed as they awaited shearing.

few buildings and homes were situated in the 200-yard strip of land between the highway and the railroad track. Near the station was a large cylindrical, wooden, elevated water storage tank, painted "railroad red," and just south of it stood the 160-foot-long corrugated steel warehouse owned by the Deseret Live Stock Company. Early residents had thoughtfully planted shade trees, which now blessed this hamlet with several large old silver poplars, elms, black willows, and weeping willows. On south, alongside the track, was a large set of corrals for shipping cattle, horses, and sheep and, just beyond the corrals, a cluster of four homes for the U.P. railroad section crew and their families. Will asked if I had pointed out the federal building to my parents and Kathryn when we came through here earlier that day. I had not. He told me to pull to a stop as we took a left turn off the highway. The "federal building" was the U.S. Post Office: five feet long, five feet wide, and eight feet tall at the gable. Tourists might mistake it for a well-built outhouse. The mail had already been picked up.

Kathryn and I had been told we would not need an automobile at the Deseret Live Stock Company, so my parents transported us to the shearing corral. They were impressed with the outfit, looked it over for a couple of hours, then drove back home to Manti, down in central Utah.

The well-traveled dirt road to the shearing corral led west past the north side of the yellow railroad station, where the elevation sign said 6,800 feet.

At the shearing corral and sheds twenty-two shearers worked in
a long line, using power-driven equipment.

On west across the railroad tracks, we stopped to open the large iron gate
with a sign on the front, "Deseret Live Stock Company." This spring range
was gently rolling, green-grass-covered hills. Going the three miles from the
railroad depot to the sheep headquarters, the dirt road wound up over two
high ridges, which gave a view of four separate herds of sheep grazing in the
distance. As we crossed over the second creek, I told Will that my dad said
this was the most feed and water he had ever seen on a spring range. It was a
great piece of rangeland.

On top of the last rise near the headquarters, it was easy to see why it was
called the "shearing corral." The shearing shed was way in the background, but
it stuck out as the major feature. It was a one-and-one-half story, Australian-
type frame building, 150 feet long. Attached to it was a corrugated-steel-
roofed shed that could house 1,500 sheep on their way into the shearing shed.
The first structures we reached were a row of five small, unpainted houses
and utility buildings. At the east end of the row was the two-room house our
family had moved into. The corrugated-steel bunkhouse was on the west end
of the row and nearest the shearing corral. Mr. Dansie had used a cabin about
in the middle, which was now taken over by Will Sorensen. A good-sized,
twenty-by-twenty-foot commissary building for storing extra groceries and
other supplies was between his house and ours. The old log cookhouse, with
an attached dining room that could seat sixty men, was just to the north of our

house, and the outside toilet was fifty yards east of the cookhouse. Another outhouse was behind the shearing corral. These buildings were not fancy, but they were adequate and suited their purpose. The source of culinary water was the well a few yards north of the cookhouse. Whoever wanted water dipped it out of the well a bucket at a time and carried it to his abode.

This was the headquarters for only May and June—during lambing, docking, and shearing—and again during October and early November, when the sheep were being worked in the fall and shipped west to the winter range. Will told me that lambing had started May 10 and was practically finished. All the docking of lambs was done, except for a herd with the youngest lambs. Shearing had started that day, June 13, with nine hundred rams processed. The shearers finished in the early afternoon, which was why we had time to tour the ranch. Will counseled me to always shear rams the first day. After shearing the rams, the ewes seemed easy for the shearers.

The crew of twenty-two shearers, the shearing crew manager, and three fleece tiers, all from Utah, had arrived a couple of days previously. Their tents and trailers were set up in a long row along the pasture fence, which ran south from the far end of the shearing shed. Three of the shearers brought their wives and young children for the nearly three weeks of shearing. Although there were no trees within a mile of the shearing corral, this grass-covered country, with a nearby lake, provided a good time for the families, and served as a rather economical vacation. The shearers ate at the company dining room, but their families were responsible for their own meals. They enjoyed this ranch setting with its far horizons, and everything clean and fresh.

Kathryn had been making the house livable and was carrying two buckets of water from the well just as I returned from the ride with Will. She reached the house ahead of me, and as I entered she told me that when Will turned this house over to us, she thanked him for having it clean and tidy so we could move in. But she had mopped the linoleum floor again, washed the cupboard shelves and all the dishes, and could now relax, knowing everything really was "Kathryn clean." This was where Will and Margaret had lived, spring and fall, for many years. He moved out so that we could move in.

The front room had a cookstove, a good-sized dining table with eight chairs, a cupboard with plenty of dishes, a sink, and a wash stand. There was one double bed in the bedroom. Kathryn's dad, Soren O. Sorensen, from Manti, gave us a three-quarter-sized folding bed for the kids, which fit very well in the bedroom. We brought our own blankets, sheets, quilts, pillows, and towels and also used some blankets provided by the company. The company supplied all the bedding in all the sheep camps. It was a well-furnished outfit.

Clarence and Marie Rasmussen. They met at the
shearing corral when she came to help the two lady
cooks at shearing time. They fell in love, married, and
raised a fine family.

It was understood that our family would eat supper that first night with
the crew over in the cookhouse. We had already become acquainted with the
two middle-aged lady cooks and the pretty, brown-haired high school girl
who helped them, all three attired in fresh, crisp housedresses. The company
had a well-understood policy regarding mealtimes. In the morning, the first
bell rang at six o'clock to arouse the late sleepers, and to let everyone know
breakfast would be served in thirty minutes. Usually, the men started toward
the eating house just before the second bell rang at six thirty. Dinner was
served at noon, with the first and only bell rung then. Supper was announced
the same way; the first and only bell sounded at six.

As the shearing crew and other men working for the company came up to the dining room that evening and sat around on the grass while waiting for the bell, I visited with nearly everyone. Many of them I recognized from three years earlier. The boys who had helped with docking, and later would work on the fence crew during the summer, were from North Sanpete High School in central Utah. Their home towns were Spring City, Mt. Pleasant, and Fairview. Will Sorensen and his late wife, Margaret, grew up in Spring City, and owned a comfortable home there. It was no surprise that several good herders, camptenders, and temporary summer help came from there. There were forty-six people at the two long dinner tables that night.

The food was excellent. Plenty of roast beef, potatoes and gravy, salads, vegetables, canned fruit, and homemade pie. Mr. Dansie had told me that he believed in hiring good people and feeding them well. Nothing fancy, but plenty of wholesome food.

When supper was finished, there were still two hours of daylight. Will Sorensen told me that there were two pickup trucks on the sheep end of this outfit, both well-used Chevrolets. I was to take the black one, as it was a little newer, and he would use the green one. He wanted the two of us to go for another ride, as he needed to show me something. We drove out north five miles to where the rolling hills of Wahsatch break off into McKay Creek, then proceeded east on the ridge on the south side of McKay. Down on McKay Creek and on to the north, in lush green grass, were about one hundred Hereford cows with their calves, a scene of ranching at its best.

Will told me there was a poison spring on the north slope, and that I shouldn't let any livestock or any of the men drink the water. One year, in late April, when the sheep were shipped from the desert, they unloaded a herd of sheep off the railroad at Wahsatch, and the herd bedded down for the night close to this spring. Next morning they had two hundred dead sheep from drinking this poisoned water. I stopped the pickup in the road, and we walked north a hundred yards to the spring. It was not fenced off, so I asked Will why he hadn't put a fence around it to keep the animals from drinking. He said that he just told the herders and camptenders not to use it.

Then I asked what kind of feed was available on the range between this area and the railroad during the year he lost the two hundred sheep. He told me the range was mostly covered with snow, and the plant called "death camas" was sticking up through the moist ground. The grass had just started. He had to get all the sheep out here into some brushy country so they could get something to eat. I could see deer tracks at the spring, and there was cattle manure in the grass a few yards below the spring. However, there were no dead animals. Death camas is the common name for the plant

The large herd of sheep in the background corral is waiting to go through the separating chute before shearing.

Zygadenus, a member of the lily family; its roots are poisonous to cattle and sheep. I told Will that the sheep might have died from eating death camas. They bedded down here for the night, but eating a poisonous plant during the day could have caused them to die during the night. He was sure that they drank the spring water and that was what killed them. I then suggested that we take a sample of water and have it analyzed. He told me that was not necessary, and that I should just keep the men and sheep away from the spring. Then he said, "You're taking on a mountain of responsibility. Handle it any way you want."

As we headed back to the shearing corral, Will told me that we would shear a herd of 1,200 ewes the next morning, and another one in the afternoon. Of course, we had to separate the lambs from their mothers while the ewes were shorn. But it still worked best to lamb at Wahsatch in May, and start shearing about June 12 or 13. He said they had tried lambing in Skull Valley in March and April, but the current system really worked best for the company. I would have to be up at daylight in the morning, and rouse the boys out of the bunkhouse to help with the herd coming in.

When I entered our house, I told Kathryn about the spring of water that Will said was poison, and which I didn't believe. I added that I would have a sample of it analyzed one of these days, so we would know for sure what the

A herd of lambs and ewes in the shearing corral.

score was. Later that evening we supplied our house with food from the commissary building. Our family was to do our own cooking and eat at our house, although it was understood we could eat with the crew at any time. Also, Will Sorensen was to eat his meals with us, except when inconvenient, when he would then eat with the crew.

The next morning a herd of ewes and lambs reached the corral just after daylight. Two sorting gates, called "dodge gates," were located ten feet apart on one sidewall of the long narrow alley called a "chute." From the corral, sheep were funneled into the chute, which was just wide enough to allow sheep to travel single file for easy separating. Will and I each manned a dodge gate, deftly opening and closing it as lambs were allowed to exit out the side of the chute into a holding corral, while their mothers proceeded through the chute and on into the shearing shed. We had about one-half of the herd separated when the first bell rang at six o'clock. We immediately closed all the gates and headed for our houses to get washed up for breakfast.

I told Will that the new chute with the wooden plank floor was the best I'd ever seen, and it prevented a lot of dust. He responded that we needed a good one. By the time we got all the sheep through in the spring and worked again in the fall, over a hundred thousand would walk through, so the floor prevented the bottom of the chute from becoming a trench. When he built the chute, he left a two-inch open space at the bottom of the sidewalls, so the dirt could be pushed out by the sheep moving through. The railroad ties under the chute kept it from rotting out.

As we neared Will's house, he told me that we had time right then, before breakfast, for him to turn the books over to me, and that I should step into his house. First he gave me the payroll book, a list of over sixty employees on the sheep operation, not counting the shearers. If they started that month, it showed the day they started work; some would leave after shearing. I had to turn in the original copy to the company office at the end of each month, and I kept a carbon copy in the book. If a man left during the month, I filled out a small card showing the number of days worked so he could take it to the company office in Salt Lake City and get paid. Monthly payroll checks from the company office would be sent to me each month, and I was to get them to all of our men.

Will then told me each herd had to be counted each month. I had to send a monthly report to Walter Dansie, showing how many sheep were in each herd; also, I had to add the herds up and show the total number of sheep on the outfit. He turned this report book over to me, along with a stack of buff-colored cards about the size of a business envelope. He showed me that the backs of these cards had thirty-one spaces, one for each day of the month. The herder ws supposed to count his "blacks" every two or three days, and record it on that date. He was also to record any death losses. The front of the card showed the number of sheep in the herd at the beginning and end of each month. The herder had to sign on a line on the card. Will closed this session by telling me that Bill Green was our range rider on the north half of the outfit, and Joe Manzaneros on the south. Their jobs were to be foremen under me. They were to do the counting of the herds out on the range and turn the signed cards in to me. It usually worked best to give the monthly paychecks to the riders, and they would give the checks to the men.

I thanked Will and was about to leave when he told me he would turn this outfit over to me and would get out of there the next week if I wanted him to, or he would stay through shearing. I told him that he couldn't just dump this outfit on me and leave. I didn't know the range—summer or winter. I wanted him to stay for a year and show me how to manage year round. He replied that he could stay, but he didn't want to be in my way.

Then it was time to get ready for breakfast. Will stood up, removed his Stetson, and hung it on a nail by the door. He removed his eyeglasses and wiped them with a hand towel. His thin white hair was damp with sweat, but before washing, he removed the heavy, herring-bone-striped, light blue coveralls that he wore over dark green trousers and a matching shirt. He always wore flat-heeled, "common sense" boots, and was the best dressed man at mealtime.

SHEARING

2

BREAKFAST AT THE SHEARING CORRAL was substantial. Either ham or bacon and eggs were served, along with hash brown potatoes, hotcakes or hot biscuits, cooked or cold cereal, and stewed prunes or fresh-canned fruit such as peaches, pears, or apricots. The pretty high school girl kept the bowls and platters filled, replenished the syrup and canned jam, and made sure everyone got filled up to last until noon.

Our crew had the herd completely separated before the shearing started at 8:00 a.m. Will told me that while the lambs were in a corral by themselves, we had to carefully look through the bunch to see if there were any long-tailed lambs that had not been docked. We needed to castrate male lambs, and cut off the tail of any lamb that was missed at docking time. When that was done, I counted the lambs as they were allowed to go through a gate into a corral, where their shorn mothers would join them. It was amazing how 1,200 ewes and their 1,300 to 1,400 lambs were able to almost immediately identify each other after having been separated for a few hours. Usually twin lambs stayed together and greeted their mother at the same moment.

Under the holding shed were two long alleys used for crowding sheep into the far end of the shearing shed. Inside the shearing shed was a holding alley five feet wide, running the whole 150-foot length of the shed. The twenty-two individual pens that held the unshorn sheep for each shearer were five feet square, and were filled with sheep from the holding alley. These pens and the alley had slotted floors to keep them clean, with the dung and urine falling through the floor to the ground. Two wranglers kept the individual pens filled, so that a shearer always had wooled sheep in his pen waiting to be shorn.

Out on the shearing floor, which was made of tongue-and-groove flooring, the twenty-two shearers worked in a long line, under a power-driven shaft

Author's Collection
The shearing crew at the large dining room, waiting to go in for breakfast.

that supplied power to their clippers. There was room for the three fleece tiers to walk up and down the floor, tying fleeces and tossing them onto the overhead conveyor belt. Each shearer reached through swinging doors to grab unshorn sheep from his own pen and pull them out onto his shearing space. Will and I walked together down the length of the shearing floor, stopping occasionally for a word with some of the shearers. The ewes had been on luxurious feed and were fat, and one shearer's response was "fat sheep make easy shearing."

After shearing each sheep, the shearer pushed it through an opening in the side of the shed, and the animal slid down a three-foot ramp into that shearer's individual pen of shorn sheep. Each of these twenty-two individual pens had room for about twenty-five sheep. Will counseled me that one person, the tally-master, was assigned to empty the pens about every forty-five minutes. He counted the sheep out of each pen, and recorded the number on the tally sheet tacked to each individual pen. That was how each shearer's tally was kept. At the end of the day, the shearer examined and signed his tally sheet. Naturally, each shearer got paid according to the count on his individual daily tally sheet.

The overhead conveyor belt transported the freshly shorn and tied fleeces to the far end of the shearing shed, where the wool grader examined each fleece as it came off the conveyor and tossed it into the proper sack. Four sacks were suspended from this overhead wool grading floor. One man tramped twenty-five to thirty fleeces in each wool sack or bag, and as each bag was

Shorn ewes exiting the shearing pens.

filled, the tramper deftly stitched the top shut with a long sack needle and white cotton twine. The full sack would be lowered to the floor below and weighed on a scale. The weighmaster recorded everything in a book and, in black ink on each sack, wrote the weight, grade of wool, and ownership brand. Each bag of wool weighed about 275 pounds, and was rolled out of the shearing shed onto a loading platform. The shearing crew could shear from 2,400 to 2,600 sheep each day, so 85 to 90 bags of wool were shorn each day. Ed Sprinkle from Boston had graded the wool for several years. At noon he told me that when I got ahead of my work, I should come and grade the fleeces while he relaxed a few minutes, just like I did three years ago. Back then he told me this was the best shearing plant in the U.S., and it still was. He was glad I was going to take over from Will.

Three men—that is, one man and two high school boys—constituted the wool-hauling crew. These three used the only large truck to haul wool three miles to the depot. There, they loaded the sacks into a railroad car for later shipment to Boston. About 40,000 pounds of wool would be crammed into each railroad car. The wool hauling crew kept busy.

One evening I took an empty quart jar and drove out to the "poison spring" to get a sample of water to be analyzed. Near the spring, a camptender was coming from the spring with a five-gallon can of water slung from the saddle horn. I told him that Will Sorensen insisted to me that the water in that spring was poison. The camptender grinned and told me that he had worked here two years, drinking the water, and no one had ever told him it was poison. He went on to say he had seen deer and cattle and horses drink there, and none had died. I knew Will was just trying to keep me informed, but the problem was not the water. Next day, Mr. Dansie took the sample of water to Salt Lake City. The report came back: "high quality, a little 'hard,' suitable for drinking."

During the long June evenings, the high school boys played softball in the short-grass area of the horse pasture just north of the houses. The shearers marveled at where the extra energy came from. Most evenings about dusk, the men gathered around a campfire near the cookhouse and swapped stories of earlier years and famous shearers. In 1929, Kathryn's father had been the winner of the World's Sheep Shearing Contest in Great Falls, Montana, sponsored by the Sunbeam Flexible Shaft Corporation. He had won $500 in gold. Several shearers had known him and held him in high regard. He had first herded sheep for the Deseret Live Stock Company when he was fifteen years old. As he matured, he became a professional shearer, and sheared at the company for several years. He had a legendary reputation.

Will Sorensen told me that Kathryn's father, Soren, could consistently shear over two hundred head of mature sheep each day, day in and day out. Shearing is back-breaking work. In addition to skill and strength, it takes determination. Soren Sorensen, during his prime, was considered to be the best in the profession in the United States. Several of the shearers would shear partially suspended in a sling, which was designed to take part of the shearer's own weight off his back. The sling consisted of a saddle cinch going under the shearer's chest, with the cinch held up by counterweights suspended from overhead pulleys. Kathryn said she was not sure when the sling was first used, but her dad and others were using it in the 1930s. She enjoyed talking to some of the shearers who knew her father.

Probably the least romantic job, or process, in the sheep and cattle business is the annual procedure of "mouthing," that is, examining the front teeth, or

Shearers at work. Several are using types of slings, evident by pulleys in
background and frame around man at left.

incisors, of each cow and ewe. Grazing animals depend on these teeth for bit-
ing off and harvesting the necessary forage. The teeth are an accurate indica-
tor of age. A set of small individual teeth worn down by constant use indicates
advanced age. In contrast to horses, sheep and cattle have incisor teeth only on
the lower jaw. Of course, they have molar teeth on both their upper and lower
jaws, but the front part of the upper jaw only has a pad, which the lower incisor
teeth bite against. Mouthing is a simple operation. It takes only a few seconds
per animal for an experienced mouther to grab the muzzle of a sheep with one
hand, spread the lips apart with his thumb and forefinger, and immediately
evaluate the condition of the teeth. With cattle, it is necessary to confine the
animal in a squeeze-chute, and use both hands to pry the lips apart.

Over the years of running this outfit, Mr. Dansie and Will Sorensen had
worked out a system of mouthing all sheep in the spring of the year, immediately after shearing, while still in the corral. At that time, ewes designated
as "old" were given a special blue-marking-paint brand. This blue brand was
simply a round dot about one and one-half inches in diameter, placed just
behind the shoulder on the right side of the back. A fifteen-inch piece of wood

Austin Christofferson, left, north rider for many years, and Clarence
Rasmussen, longtime south rider, at the corner of the
meathouse at the shearing corral.

sawed from a broken shovel handle or off an old buggy spoke made an excellent brand, which was easy to use and lasted for years. Will Sorensen told me
that these old blue-dot ewes would stay in the herd with their lambs during
the summer. However, in the fall, we would mouth just the blue-dots, probably about six thousand, and keep only those that could go another year. If a
ewe is not suckling a lamb, she is called a "dry." A dry has practically no udder,
while a ewe nursing a lamb is obviously giving milk and has a large udder.
Marking freshly shorn dries with colored chalk on the back makes it easy to
separate them from the herd as they go through the mouthing chute.

From the holding corral, the sheep were driven into the long, narrow
mouthing chute, which was three feet wide and one hundred feet long. The
sides were three feet high, constructed of lumber. About seventy-five ewes,
packed tightly against each other, filled the chute. The closed gate at the
rear of the bunch prevented the sheep from moving while being mouthed.
Usually two men did the mouthing, working from outside the chute, one
man on each side. Austin Christofferson and I did most of the mouthing.
He was Will Sorensen's nephew, and had practically grown up on the outfit,
starting as a camptender when he was only twelve years old. He was "forty-
ish," still a young man, five feet six inches tall, with a head of dark hair and a

wiry build, having spent many years in the saddle as the north rider. Austin had a keen sense of humor, which I enjoyed as we worked across the chute from each other. He now worked on the so-called headquarters crew, filling in where needed.

Working along the mouthing chute, we were closely followed by the man applying the blue dot. The mouthers would say "brand this one," or "blue dot," and the paint was immediately applied. Usually the herder had the responsibility of applying the fresh new brand designating his individual herd of sheep, so he followed along behind the blue-dot brander. He made sure not to brand anything carrying the chalk mark that designated a ewe as "dry." Will told me that if a ewe was old and didn't have a lamb, there was no need to keep her. So these were classified as culls, given an extra chalk mark on the back, and cut into a separate corral. The cull ewes removed from a herd had to be counted, then driven to a holding pasture adjoining the shearing corral. Later, when shearing was completed in early July, these culls would be shipped to Omaha, Nebraska. There were two large holding pastures, one for the culls, and one for young dry ewes that would be kept.

The Deseret Live Stock Company, and almost all range sheep operators, kept two black sheep for about every hundred white sheep. Black sheep were used as "markers," because they were easy to see in a herd of white sheep. Herders were expected to count their blacks every day if they were moving to a new location. If they were missing a black sheep, it was likely to be in a bunch of white sheep that were also lost. The herders would know approximately where the sheep grazed the day before and would go there to look for the lost sheep. At shearing time, the black sheep in each herd were the last to be shorn, because the black wool had to be sacked separately. After the blacks were shorn, the shearing floor was swept clean with brooms, so no black fibers would get mixed in with the white wool. Black wool does not take dye when wool is being manufactured into fabric. Therefore, it had to be kept separate at shearing.

Will and I agreed that running a large livestock outfit was a business of counting and recording. After each herd was shorn, the tally recorded on each shearer's pen was totaled for all twenty-two shearers, and this number needed to match my count as the sheep left the mouthing chute. I pulled a durable, pocket-sized notebook from my shirt and told Will that this was for keeping track of the many small bunches that we handled in each herd. My counting was carefully done and recorded, and the day's end tallies came out exactly the same from the shearing pens and from the mouthing chute.

As each herd was shorn, I gave the herder a new card showing his name, herd brand, and the numbers of ewes, lambs, and blacks in his herd. He was told which summer allotment he was to graze—most herders went to a range where they had previously summered —and the route his herd was to take in order to get there. He was also told which herd of sheep was ahead of him, which herd would be following, and about how far they should travel each day as they grazed on west to the higher summer range.

Earlier during the lambing process, ewes with their lambs were separated from those that had not lambed. When the lambing season was finished, those ewes that did not have a lamb were gathered into "the residue," mature dry ewes that either had not lambed or had lost their lambs. When this particular residue herd came in to be shorn, considerable time was spent looking them over while they were still in full fleece. Will and I agreed that mature ewes that had not lambed that year and were dry the previous year should be put in the culls to go to Omaha. So, prior to shearing, if a ewe came in to be shorn and had no lamb, and was also wearing a "dry" brand from a year ago, she was marked on the nose with colored chalk so she could be identified after she was shorn. Young ewes would be given another chance. However, those four years old or older were culls.

Kathryn had to keep close track of Bill and Diane, but they could play in the fenced-in area, which included our house, the cookhouse, commissary building, woodpile, coal shed, and covered well. When she was caught up with her work, she took the children by the hand and walked over to the shearing shed and corrals. They did not enter, but watched from outside. As the herders and camptenders gazed at her and Bill and Diane, I was proud to let them know that the good-looking young woman and two children were my wife and family. I told Will that Kathryn and I went together during my senior year at Manti High School. She was a year behind me. I went to college at Utah State in Logan that fall, and she came there a year later. I was a junior and she was a sophomore when we got married in February 1941. The following December, the Japanese bombed Pearl Harbor. In the U.S. military draft, I was classified as "3A" on account of being a married man. Our son Bill was born in September 1942, and Diane was born in October 1943. We were all glad to be here.

That evening, in the privacy of our home, Kathryn told me that when she and the kids walked over to the corrals, she was surprised to see a pressurized water system with hoses to sprinkle the corrals, in order to keep the dust down. She said the sheep got water piped into the corrals, but we humans had to carry water from the well. I told her that after we'd been here awhile, I'd learn a few things, but I couldn't agitate about getting water piped into the houses until I knew the whole situation. We were going to be on the summer range in three weeks, and Will said water was piped into our house up there.

At the Deseret Live Stock Company, the ewes lambed for the first time at two years of age. Will Sorensen told me that we usually had 7,500 yearling replacement ewes, in two herds of 3,750 each. These were kept by themselves during lambing. During shearing, we often put 1,500 yearlings under the shed in the shearing corral at night if it looked like it was going to rain. We knew we couldn't shear wet sheep, so if it rained, there would be enough dry sheep to keep the shearers going for at least half a day. Will told me to combine young dry ewes from the holding pasture with enough yearlings to make summer herds of 3,000 to 3,300. We put these young ewes single file through the narrow branding chute, gave all 3,000 the same brand, and sent them to their summer range. A herd of dries went out every week. These dry herds were summered in good feed, but in the lower and less luxuriant part of the summer range. Ewes with their lambs were given the best of the high country.

Shearing was a great time at the Deseret Live Stock Company. Many visitors came to just watch the magnitude of the operation, and see the vast herds of sheep moving in and out. Young women visitors had a magical effect on the young men working in the corrals. Instead of walking to open a gate, they ran. Instead of climbing over a fence, they simply put a hand on the top rail and vaulted over. If asked a question, they were a fountain of information. I told Will that we ought to have that pretty high school girl who helped at the cookhouse walk out here to the corral in the middle of the afternoon. That would be like hiring a fresh crew of high school boys. Will said that we were all like that, and the right woman could bring out the best in a man. He turned away from me, removed his glasses, and wiped his eyes. As he turned back, I glanced out across the corral. Will said in a soft, almost quivering voice that when you lost your mate, you had some hard days in front of you. I told Will I knew it was a tough time. That evening, I confided in Kathryn that Will was still feeling pretty rough about losing Margaret, and that I felt sorry for him. It was too bad they didn't have any kids. She replied that Will was a very considerate man. Some people said he didn't like kids, but he was thoughtful and kind with ours. He was almost like another grandfather. We agreed that Will Sorensen was a mighty fine man.

On June 20, the cowboys from the Home Ranch distributed over one hundred horned Hereford bulls into bunches of cows and calves, which were spread all across the range between headquarters and the shearing corral, a distance of twenty miles. By the time they reached the shearing corral in the late afternoon, there were just two cowboys driving thirty bulls. These bulls were put into the cattle corral for the night, and would be distributed on south the next day. Their two saddle horses were stalled in the horse barn. Although these were the shearing headquarters, there were facilities

Loading bags of wool to be taken by truck to railroad
car at Wahsatch, Utah.

for cattle and horses. When shearing was done for the day, I walked over to the cattle corral to see the bulls and chat with the cowboys. Just then Ralph Moss, manager of the Home Ranch, drove up in his pickup to take the cowboys back home for the night. We recognized each other from when I had been there three years earlier, and we had a pleasant visit. I told him that on my trips out across the range, I had stopped to admire all those Hereford cows and calves, and that the big, well-muscled bulls were doing a good job. Ralph replied that the men at Deseret preferred the Herefords. They respected each other's horns, and had a natural tendency to spread out across the range away from each other. They were good travelers and had enough hair to withstand hard winters.

When I asked Ralph how many bulls they used per hundred cows, he said one bull for twenty-five cows. This was an easy range, of gently rolling hills with plenty of grass and water, and the cattle didn't go up onto the high summer sheep range. I knew cattle have an average gestation period of 282 days, and commented that calves ought to come during the last three or four days of March. He told me that was when calving started, but that they liked to get most of the calves in April. That was early enough for cows calving out in the open in this high, cold country. The cows and calves went out on the range about May 15, depending on how the feed was. Calves were branded at the roundup in the fall. He told me to keep my eyes open as I got around the range, and let him know if I saw any problems. I informed Ralph that I was interested in the cattle just as much as I was the sheep. He was pleased. We visited a few more minutes, and then he said they needed to go back to the ranch. He would send us a quarter of beef when the boys came back in the morning. I thanked him.

When they came the next morning, they hung the beef in the meathouse, saddled up, and drove the bulls on south. As they were leaving, I told Will that those cowboys rode good horses, and asked if they were all raised on the Home Ranch. He replied that they were all raised here. The horse brand looked like this:

$$\overset{|}{X}$$

and was on the left hip. I said, "That I over X makes a good horse brand." Will answered, "You can call it I over X if you want, but Ralph Moss and I call it the 'cross-eye.' It's an old brand." The cattle brand was a capital J with a quarter circle above it:

$$\overset{\frown}{\text{J}}$$

and was on the right side of the animals. The brands were legible but not burned too deep. It was easy to see that the branding of cattle and horses was done with care and expertise. I had branded a few calves and yearling foals at home and could recognize a good job.

When shearing was completed the evening of July 2, nearly 40,000 sheep had been shorn and 1,500 bags of wool shipped. The wool weighed nearly 400,000 pounds. Later, the international wool trade reported that the Deseret Live Stock Company produced the largest clip of wool in the world from an individual ranch. These almost unbelievable figures, and the volume of the operation, just added to an underlying pride in the Deseret Live Stock Company. Kathryn and I were, frankly, quite proud to be part of this outfit.

Mr. Dansie was at the shearing corral and wrote individual checks to each shearer and to all temporary employees. Then he and I had a conference. When he asked how many cull sheep we had going to Omaha, I told him 640 total, which included 80 old rams and 560 ewes. He said he would order three double-deck rail cars, so we could put the rams in one deck, and the ewes in five decks, and that I should go over to the depot that evening and see what time we could load the next day. I told him all the summer herds were heading

Keith Moss McMurrin Collections
Bags of wool ready to be loaded in a railroad box car.

for the mountains, so all we had to handle was the bunch of culls. He complimented me on how smoothly the work had gone during shearing. I told him Will Sorensen was teaching me, and that I would be at a terrible disadvantage without him; Will and I were able to work together.

The next afternoon, when the cull sheep were securely on the Union Pacific Railroad en route to Omaha, our small crew returned to the shearing corral. Tom Judd, the company trapper, had agreed to cook for a couple of days. Kathryn helped prepare supper. That evening, Kathryn and I took young Bill and Diane into Evanston, Wyoming, so they could enjoy part of the booming celebration getting underway for the Fourth of July. Things really started on the third and ended on the fifth. As we drove into town, I said, "We haven't seen any electric lights since we arrived at the shearing corral three weeks ago. I hope all these bright lights don't throw us into a molt." Bill asked, "What's a molt?" Kathryn answered, "It's when chickens go through a change in the seasons, and their old feathers fall off their bodies. Nothing like that will happen to us. How would you like to start this little celebration with some ice cream?"

Tom Judd, Deseret Live Stock Company trapper, delivering a load of coyote
pelts to the R. C. Elliot Company in Salt Lake City (circa 1940).

We enjoyed the sights of Evanston about all we could stand for one night. It
was eleven o'clock when we returned to the quietness and isolation of our house
at the shearing corral for a night of complete rest. Walking from the pickup
toward the house, we stopped to appreciate the night sounds of croaking frogs
in the pasture and the soft nicker of a horse just over the fence. The night was
clear and the stars big and bright. I pointed out the Big Dipper, the North Star,
and the Little Dipper. Kathryn said she knew I was enjoying this, but that the
kids needed to go to bed. While those three visited the outside toilet, I lit up the
gas lantern in the house. It was obvious Kathryn did not enjoy this rugged way
of living. However, we could handle it as we looked to the future.

Now and then, I reflected on which men on the sheep outfit had their
hopes of succeeding Will Sorensen as sheep foreman dashed to pieces when
I came on the job. Everyone was cooperative, but there were three, or possibly four, who must have been disappointed. Kathryn and I had agreed that
my so-called "advanced education" was never to be mentioned. However, my
growing-up years herding sheep, riding horses, and working with cattle on
our family ranch were a great asset.

3

The High and Glorious
Summer Range

With shearing completed, it was time to close up the houses at the shearing corral and move to summer headquarters in the high country. The sheep operation was responsible for a summer fencing crew of six husky high-school-age boys, a man cook, and Bill Watts as fence crew foreman. Bill had worked for the company for several years before going off to fight in the U.S. Army in the Pacific area in World War II. He was in his early thirties, unmarried, and was a man of great energy and ability. He was muscular and trim, five feet ten, and had a great sense of humor.

Will Sorensen and I ate breakfast with these men that last morning at the shearing corral, so the day's work could be outlined. The fence crew would camp high on the mountain a couple of miles from headquarters. I told this group that we needed two volunteers to ride saddle horses and drive the extra horses to the summer range. All six boys volunteered, and two were called upon. After breakfast, I helped bring the horses from the pasture into the corral and got them saddled up. I told the boys that it was twenty-eight miles, mostly uphill, so they shouldn't push too hard; a steady walk was about right. Will indicated that he would oversee getting all the cases of food segregated into a pile for the fence crew and one for headquarters, and that the boys could help him before they loaded the trucks.

Moving day was not the most pleasant. Kathryn and I loaded all our food, bedding, and personal belongings into the pickup truck, covered the load with a canvas tarp, and put the kids in the seat with us. Will Sorensen drove his personal black Chevrolet coupe at the head of the procession, followed by Austin Christofferson in the big truck, towing his trailer house. Next, Tom Judd

Dean and Kathryn Frischknecht at the sheep summer range headquarters, with children "Bill" Dean (age 3 years 9 months) and Diane (age 2 years 8 months), in early July 1946.

drove his personal vintage pickup, followed by Bill Watts in the green pickup, and our family in the black one. We climbed in a northwesterly direction to the head of Lost Creek, about ten miles south of Monte Cristo, a landmark peak. Although we traveled a winding, dusty, dirt road, it was exhilarating to be on the high range, at about a 9,000-foot elevation. Grass and wildflowers were luxuriant. The quaking aspens were green with fluttering leaves, and firs, pines, and spruce gave off that distinctive fresh, piney fragrance of the high country. I told Kathryn and our kids that this seemed kinda' close to heaven, no matter if it was isolated.

The largest house at headquarters was an old, rustic, two-room log cabin, with a green lawn front and back. Will Sorensen told me that he built this cabin a few years ago for Margaret and himself. The largest room had a front door and a back door, and served as kitchen, dining room, and living room. The other room was just right for two beds. This house had a corrugated-steel roof that glistened in the sunlight. Bill and Diane called it "our big silver home." Will Sorensen pointed out the modest one-room cabin he would use, and a fairly large one-room cabin surrounded by quaking aspens, which was reserved for Mr. Dansie. Austin Christofferson parked his trailer a few yards from Will's cabin.

Will told Kathryn and me that water was piped into our house from the spring up the draw. Of course, the pipes were drained for the winter. He told Kathryn to turn on the kitchen faucet, and he would help me get the water hooked up. It took a while to get the water flowing, and it ran for several minutes before Kathryn dared use it. When Will and I got back down to the house, she said that this running water would be a pleasure compared to the shearing corral, where we carried water in buckets from the well to the house. These houses took on a musty odor while closed up from October to July, so Kathryn kept both doors open while she cleaned the dusty cupboard shelves and swept the floor before moving in.

Will Sorensen and his late wife, Margaret, had worked hard to make this a homey, livable place. They had nurtured the flowering vines growing on the sides of the cabin, and had hauled loads of topsoil for the lawn. Later, I mowed the lawns and Kathryn irrigated them with a garden hose attached to the water system. To the west of the cabins, a large, open-front log shed served as the garage, and could house three trucks or autos parked side by side. On the south end of the shed was a closed room with a wooden floor, where extra food and supplies were stored. Will told me that I would have to keep a padlock on this door, so people wouldn't just come and help themselves. On the north end of the shed was the tack room for saddles and horse equipment, and for two fifty-gallon metal barrels where oats were kept safe from squirrels,

Clarence Rasmussen hauling a load of sawn logs to the end of the
fence on the summer range.

chipmunks, and birds. Will said his experience was that the extra horses would
stick pretty close to headquarters if they got a small feed of oats when they
came in each day, and oats made them easy to catch. I told Will that I was
happy we had six good horses around here this summer; he and I would do a
lot of riding as he showed me the outside boundary of this ranch. He indicated
that it would take several long days, but we would get it done.

The next morning, along about eleven o'clock, after the sheep had "shaded
up" under the trees during the heat of the day, Eldon Larsen rode in from his
camp near his herd on Dog Pen Ridge, about four miles south. He told me
that if we would get our family into the pickup and ride out to his camp, he
would show us where a cougar had killed a buck deer. He would leave his horse
here at headquarters, and ride out and back with us. We visited the area, situated in a grove of aspens about thirty feet from a small watering pond. Eldon
reasoned that the buck had been drinking, or had just finished, when it was
attacked by the cougar. The partially eaten carcass was well concealed, with
small branches and soil scraped over it. However, Eldon's dogs had smelled
the carcass, which alerted him to go investigate. This two-point buck was still
in "velvet," that is, his horns had not shed the soft outer covering that pro-
tects the growing horn. During late summer, buck deer remove this somewhat
skinlike outer covering by rubbing their horns against trees or woody shrubs.
The true antler is then exposed. Eldon figured the cougar must have killed the
deer just the day before his herd of sheep arrived. It was possible the cougar

Author's Collection

The fence-building crew riding back to their camp.

would not return after all of us and the dogs had been to see the dead deer. Eldon checked on the carcass each day; the cougar did not eat any more of it there, but other scavengers reduced it to bare bones.

The two riders, Bill Green and Joe Manzaneros, each had his own camp in a strategic location on his part of the range, and each had two top saddle horses so he could cover the range. The riders contacted their herders and camptenders every few days. If a herder was missing a bunch of sheep, the rider helped locate and return them to the herd. I made out a detailed monthly report on the total number of sheep and turned this in to Mr. Dansie. These reports from each herder showed areas having excessive death losses from coyote attacks, or from any other causes.

The company was in the process of building a fence completely around the outside perimeter of the 250,000-acre ranch. Cedar posts and wire were used in the lower elevations, but in the high country, the fence was constructed of logs. "Log and blocks" was the common name for such a fence. Logs used in the fence were cut from dead trees, either quaking aspens or conifers. These well-seasoned dry logs did not sag when laid horizontally on the fence. Usually logs that were about one foot in diameter were used, and were cut about sixteen feet long. The "blocks" were four feet long and were

Dean Frischknecht at the summer range,
with brook trout caught in nearby
beaver-dam ponds.

set at right angles to the end of the log. The fence was three logs high. Each bottom corner of the block was set on a flat rock, to keep the wooden logs off the ground and prevent decay.

The young men on the fence crew began their summer work in the high country by repairing the existing fence, and then built several miles of new fence. They worked hard, and were fed exceedingly well. We kept the cook in plenty of fresh meat, usually mutton, but sometimes their meat was wild, courtesy of Will. Occasionally, the boys would go fishing on Sunday, as that was a day of rest and relaxation. They were too far away from town to attend church, so fishing was fun, and fish were plentiful in the numerous beaver-dam ponds along the creeks in the bottom of the draws and canyons. Quite often during the summer, camptenders came to headquarters for mail or something else, and brought us a nice catch of brook trout. These trout were usually eight to twelve inches long, firm-fleshed and tasty. The cold mountain water was crystal clear, unless beavers were working in the area. Plenty of grasshoppers and other insects made these streams a good habitat for fish.

The company had a practical and organized system of keeping the sheep camps and ranches in food. Every six weeks a new list of needed food and supplies, called a "grub order," was made out for each camp and ranch. These orders were filled at ZCMI Wholesale Grocery in Salt Lake City. Each grub order, consisting of several boxes, was individually prepared; each box was properly tied with string, and labeled according to the name or identification number on the grub order sheet. Also, hardware items were identified with an attached tag.

Over the years, the company had developed a grub order book with over three dozen food items listed on identical sheets that were detachable from the book. The list included such items as flour, condensed milk, sugar, coffee, tea, oatmeal, rice, raisins, dried beans, potatoes, hominy, cheese, prunes, honey, jam, pickles, and several kinds of canned vegetables such as corn, peas, green beans, tomatoes, and spinach, several kinds of canned fruit including peaches, pears, apricots, and apples, and other items such as catsup, mustard, peanut butter, lard, and all kinds of spices. To make out the order for each camp, all that was necessary was to read down the list and write in the desired quantity in a blank at the side of each item. Several blank spaces followed the itemized list, so that other items could be ordered. At the top of the sheet was a space to write in the name or number of the sheep camp or ranch.

On the sheep operation, the two riders had the responsibility of helping each herder or camptender make out his grub order sheet. They then gave these to me to send to the company office in Salt Lake City. In July, soon after reaching the summer range, Will Sorensen picked up the order from the fence crew, and Kathryn and I made up the order for the sheep headquarters, which included the food for our family. Will told me to make sure I ordered plenty, enough to last six weeks. He didn't want us to run out of anything. He gave me the order for the fence crew and said it was a big order, but nothing was out of line. The crew worked hard, and they could have about whatever they wanted. I looked it over. It was mostly case lots, as the cook knew what he needed.

Will counseled me that before I sent those herders' grub orders to Salt Lake City, we needed to go over them, because sometimes it was hard to read their writing. Soon I learned that the main purpose was to strike out some items. Will told me to read down the herders' grub lists and draw a line through "jam," because they hadn't been able to get it for years, and to strike out "pickles," because they didn't need that stuff. I knew when those herders and camptenders ate with us at the shearing corral, we served plenty of jam and pickles. So I cancelled out those items on the lists from the camps. I asked,

"This one rider wants six jars of jam. The other hasn't ordered any jam. How do you want this handled?" He told me to strike out "six" and write in " three," because that man didn't need all that jam. I then told him that Kathryn and I ordered eight jars of jam for our headquarters, and he replied that that was all right, and that we should order whatever we wanted. I did as instructed, but thought this policy was not consistent, and didn't like it.

The next morning, Austin Christofferson, the truck driver, took the grub orders to Salt Lake City. He was to haul a piece of farm equipment from Salt Lake City to the ranch in Skull Valley, where he would stay overnight. The next day, he would load ten one-hundred-pound bags of potatoes from the spud cellar at the ranch, pick up the load of grub in Salt Lake City, and return to the mountain. ZCMI Wholesale Grocery could make up the grub orders in one day. As soon as he left for Salt Lake City, Will Sorensen and I drove down the mountain to the Home Ranch headquarters, near the northeast corner of the ranch, to pick up thirty fifty-pound bags of flour to be distributed on "grub day," which was scheduled for the following day. Mr. Dansie purchased flour in large quantities, a semi-truckload of twenty tons or more at a time, and had it stored in the basement of the big house at the Home Ranch.

As we were driving down the mountain, I told Will that Mr. Dansie indicated this so-called Home Ranch contained over 250,000 acres, with nearly 230,000 acres of it being privately owned. Will replied that there were a few thousand acres of federally owned land within our fence, but they were in a checkerboard pattern, mostly in two townships. Some of the private land was part of the railroad grant that the company purchased, and the federal pieces of land were those alternate sections that the government did not give to the railroad. A "section" is one square mile, 640 acres. I knew that the government gave certain railroads "every other section in alternating townships on each side of the railroad," in order to get the railroads built. So the U.S. government and the railroad each owned eighteen of the thirty-six sections in some townships. A township contains thirty-six sections and is six miles square. Will indicated that the government owned about thirty sections within our border, but they wouldn't sell them to us. He said someone came and looked at it occasionally, but that didn't cause a problem.

The Home Ranch headquarters and a lot of the surrounding country was once owned by the Neponset Land and Livestock Company. The owners lived in Boston. The Deseret Live Stock Company wanted to buy it, so in 1917, the Neponset owners sent a man out on the train. Ez Hatch, president of the Deseret Live Stock Company, was to meet him at the hotel in Evanston, Wyoming. Will told me the price was almost $200,000, a lot of money in 1917, when they made the deal. According to Will, Ez Hatch drove a team of horses pulling a bobsleigh from the company property into Evanston. Ez was wearing a sheepskin-lined overcoat, blue-denim bib overalls, and old-style, four-buckle, red rubber overshoes. The man from Boston was really dressed up, wearing the city clothes of a proper businessman, and he was terribly disappointed when he saw Ez Hatch and another man or two from the Deseret Live Stock Company. He thought he had made the trip for nothing. Finally he asked, "Just how in the world do you propose to pay for this ranch?" Ez replied, "We thought we would pay cash," and he wrote a check for the full amount. This was a tremendous addition to the company.

Mr. Dansie had told me the big white house at the Home Ranch was built in 1941, at a cost of $15,000. It was a mighty fine house. An oil company paid $20,000 to lay pipeline across company property, so Mr. Dansie and the Deseret directors decided to use that money to build a house they could be proud of. Mr. Dansie liked to have cattle buyers stay there when they were making a deal on the cattle for sale. Will and I planned to eat noon dinner there that day, and I wanted to take a tour through the house. It had a big dining room, and some of the hired men ate there. Ralph Moss always kept a good cook.

As we approached ranch headquarters, it was evident that haying season was well underway. Will told me the hay meadow was ten miles long. The hay crew would put up about ten thousand tons of meadow hay and stack it out in the meadows. That's where the cattle would winter, and it took about one and a half to two tons of hay for each cow for the winter. After the calves were weaned in the fall, they got fed separately, in large lots close to the ranch headquarters.

At the house, Will told the lady cook that we needed to get thirty bags of flour, and that we would like to be there for the noon dinner. That was fine with her. When we got the flour loaded, we were to come in and wash up. There would be fourteen of us for dinner, but there was plenty of food and plenty of room. Twenty tons of flour was a lot of flour to be stacked in the basement of one ranch house. However, it was built to store plenty of supplies. We loaded the flour, washed up, and walked into the living room. It had a high ceiling, a large fireplace made of stone, and a stairway with a banister that circled an upper balcony leading to several upstairs bedrooms. I was not disappointed. The front porch on the east side of the house was screened, and looked out across miles of hay meadows. To the north of the house, geese and ducks were floating on a large, spring-fed pond that was surrounded by big old shade trees. The spacious lawns, front and back, were well kept. This was a beautiful setting for the headquarters of a great ranch. Will told me that the hay meadows were flood irrigated, using water stored in the Neponset

Home Ranch headquarters.

Reservoir on the "little desert,"about four miles southeast of the house. I had seen that reservoir when we went down to see the dry herds during shearing. It covered about a square mile, a great setup for irrigation.

The cook rang the bell, and soon fourteen of us were seated at the long table in the dining room. We ate standard fare: very tender roast beef, potatoes and brown gravy, green salad, vegetables, hot rolls, milk or coffee, and apple pie. That was a pleasant visit.

Next day, on the summer sheep range, the north-end camptenders came to headquarters for their boxes of grub, and I met the south-end camptenders at the salt-storage house, situated at the head of Horse Ridge. It was an efficient system. During the summer, each camptender helped two or three herders. After a few days, Bill Green, the north rider, came over to chat with me. He indicated that there were some food items that were rationed during World War II, as they were just not available, but now using that

The big white house, far left, is partly hidden by trees.

excuse for not furnishing jam was aggravating some of our men. He told me that if those guys ordered six jars of jam for six weeks, we should give it to them. If we were not going to give jam to the men out in the camps, then we should make out a new printed list and leave that item off. I confided to him that I knew this was an inconsistent policy, and I didn't like it. I said that Will Sorensen wanted to hold down the cost of the grub orders, but that I thought we had to treat all employees the same with the grub policy. Mr. Dansie visited summer headquarters a couple of weeks later and wanted to see several herds of sheep and look over the range. When he and Will and I were together, it was the right time to discuss the grub orders, so I brought it up. They agreed that I should use my own judgment and handle it the best way for the company.

The two riders on the summer range did not have responsibility for the two herds in Heiner's Canyon, nor for the buck herd. Will Sorensen and I handled their grub orders, kept them in salt, and counted the herds every month. Counting out on the range was similar to sand going through an hour-glass. The sheep passed through a narrow area, with men on each side count-ing them as the sheep went by. One day Will was with me when I counted both herds. Each had over 3,300 sheep. The first herder had all his sheep accounted for. The second was missing thirty sheep. I said, "According to my count, you are out thirty sheep. Have you got all the blacks?" The second herder said yes, he counted seventy-six blacks, and that's what there were sup-posed to be. The camptender said to the herder, in Spanish, that "these guys always tell the herders that they are out a bunch of sheep, but I know you are not out any sheep." Just then a little bunch of sheep came single file over the top of a hill, about three hundred yards distant. Will said, "Here come your lost sheep." We counted twenty-seven as they came down the hill. It couldn't have happened any better. From then on, they believed my count. Other than coyote kills, losses were light.

In addition to the quarter of a million acres of mostly private range included in the Home Ranch, the company had a permit to graze three herds of about a thousand ewes each, plus their lambs, on the Cache National Forest. This area was north of the private range, but south and east of Monte Cristo. These three herds grazed on forest land for only a couple of months and then were brought back inside the boundary fence. These forest allotments were a little small for three herds, so after the summer of 1946, Will and I worked out an agreement with forest officials whereby we would graze only two herds, where there previously had been three, but we would keep those two herds on the land until mid-September. This would work much better, and the forest officials were pleased with this arrangement.

As summer came on, we made trips to the post office at Wahsatch about every week or ten days to mail letters and pick up mail. These trips were timed so that if any herders or camptenders were leaving, they could be hauled to the bus at Evanston, and usually their replacements could be picked up on the same trip. Will told me that we didn't want a truck going empty, so we should always cut a load of wood to be unloaded at the woodpile down at the shearing corral. Dry quaking aspen, pine, or fir trees were cut, trimmed, and sawn to fit the length of the truck. Later in the fall, at the shearing cor-ral, we would hook up the power saw and cut those poles into stove-length wood. We hauled plenty of wood, enough to last through the fall and the fol-lowing spring. On these trips to Wahsatch, we always went to the company warehouse and loaded the truck with bags of sheep salt to take to the summer range. Livestock need salt in their diet, and they eat more salt in summer when the feed is lush and green.

When it was time to turn in the next grub orders, Will Sorensen was gone for about ten days, so I carefully read each individual order to make sure everything was legible, and that the supplies ordered were in keeping with company policy. The policy now included modest orders of jam and pickles. On the first grub order list I examined six weeks ago, one of the older herders, George Rasmussen, had listed "one Chore Girl" in one of the blanks at the bottom of his grub order sheet. I had laughed and drawn a line through that item. Now, six weeks later, George again wrote in and underlined "one Chore Girl to clean pots and pans." Again, I just laughed, and said to Kathryn, "George Rasmussen really has a sense of humor. This is the second time he has ordered a Chore Girl." Kathryn said, "Well, get him one." I replied, "Look, Kathryn, George is just joking. He knows he can't have a girl at his camp." She explained, "Dean, what George is ordering is a Chore Girl to help clean the pots and pans. Chore Girl is the brand name of a steel-wool product that is made for this purpose. So please get George his Chore Girl." She, or "it," came on the next order. It's amazing how much I learned about specific brand names of different products. I later told this little story to an dear older friend, who told me that some people get their education without going to college, and some get it after college.

Kathryn, Bill, and Diane rode with me in the pickup as we traveled to the Home Ranch for thirty bags of flour. At the ranch, cowboys had gathered twenty young, unbroken horses off of the range. The cowboys were selecting new horses they were going to start working and training to be cow-horses. The company owned about four hundred horses, so the cowboys could be very selective. A boy about fifteen years old was with his folks from Salt Lake City, who were visiting friends at the ranch. He had climbed up on the top rail of the fence and was watching the ranch manager, Ralph Moss, deftly rope the selected horses. The boy pointed to a beautiful three-year-old, blue-gray gelding and exclaimed in a loud voice, "I would really like to ride that horse." Ralph Moss looked at him and asked, "Have you ever been on a horse?" "No," came the answer. "Well," Ralph said, "you'd be even. That horse has never had anyone on him."

Ralph Moss was one of the best horse ropers in the country, and it was a pleasure to watch him in a corral. I told Kathryn that Ralph roped a young horse for me one day, and that I'd never forget how easily he put the loop on that horse. In the round corral, Ralph got the horses milling in a counter-clockwise direction. He faced the oncoming horses, and almost backed his left shoulder into the horse he wanted to catch, so that the horse paused slightly along the fence. Then Ralph turned his body to the right almost a three-quarters turn, and the horse gave a new burst of speed, thinking it was getting away. Ralph simply put the loop smack in front of the running horse, and the horse stuck his head right into the loop. Ralph told me that that method of roping is called the "Hoolihan." Ralph was a true artist when it came to roping horses.

In a few days, Will Sorensen returned, and he brought copies of the cost of each individual grub order, both the first ones from mid-July and the second set, the September order I had approved and distributed while Will was gone. We sat down together to compare the costs of these two sets of twenty-seven individual orders. Ten were about the same. Seventeen of the second round of orders were from five to six dollars higher. Will told me that he and Walt Dansie looked these over in the company office in Salt Lake City, and Will wanted to give them to me so I could see what was happening. The total cost of food items on the second order was $100 more than the first. They wanted me to be aware of the situation, and keep those costs down. I told Will that the only way I could justify this was by noting that our men felt a little better after the second grub order was distributed. Maybe because of that, they would go the extra mile for us, and maybe we would have less employee turnover if they were happy about working for us. I knew the extra cost would be about $800 per year, but one disgruntled herder could cost us more than that. Will told me to keep costs as low as possible. I could tell from Will's tone of voice that this slightly liberalized grub policy was not to his liking.

Here is an example of a Deseret Live Stock Company grub order sheet, used in the 1930s. This was revised in 1940, with more food items listed. Note the spelling of "fence." Wagon grease, essential for axle lubrication, was listed close to "jam" so it would not be overlooked.

DESERET LIVE STOCK COMPANY

Camp No.......... By... *Fence Crew*

....193...

Item	POUNDS CANS PKGS.	Item	POUNDS CANS PKGS.
Flour	100	Tomatoes	1 Case
Sugar	25	Lard	
Coffee	8 lb	Milk	3 Cases
Tea	2 Pkg	Honey	
Raisins		Jam 5 lb buckets asst	12
Apples		Wagon Grease	
Peaches		Soap Laundry 20, Toilet 12	
Prunes		Cartridges	
Beans		Matches	2 carton
Oatmeal Mothers with China	3 Pkg	Macaroni	
Rice		Cheese	20 lbs
Baking Soda		Butter	12 ll
Baking Powder		Yeast Monel	2 dozen
Pepper		Ketchup	12 bottle
Potatoes	200	Pickles 3 qts sweet - 6 dill	
Syrup		Peanut Butter	
Salt		Broom	
Corn	1 Case	Lamp Globes	
Peas	1 Case	Cinnamon	
Allspice		Ginger	
Nutmeg		Cloves	

Powdered Sugar 15 lb , Corn flakes 10 Pkg
Jello asst 12 pkgs
Mince Meat 6 Pkg
Corn Starch 2 "
Spinach 10 Cans
1 Case eggs

4
Great Trail Drives

During that first week of September, while we were still living on the summer range, Will Sorensen and I went to the Home Ranch to get a pickup load of baled hay. Ralph Moss told us that he did not have a bale of hay on the place. The hay was all stacked loose, but he would hook up a team to the old-time baler and bale a pickup load for us. There was a team of horses harnessed and standing in the barn, and we could use them for a while before they went back out into the fields. For the first and only time in my life I saw a stationary hay baler, which was operated by a team of horses driven in a circle around the baler. A wagonload of meadow hay was parked so that the hay could be pitched from the wagon to the conveyor feeding the baler. Ralph said he would drive the team, and Will should take a fork and pitch hay right into the conveyor. The team provided the force to compact the hay into the pressure box. When we got a bale, we would fasten wires around it, and then it was forced out of the baler. My job was to stack the bales onto our truck. That old-time baler worked. I loaded twenty-four bales of grass hay, and it didn't even take an hour.

When we finished baling, Ralph Moss asked me if I could use a beautifully matched team of big, bay, four-year-old draft horses. I replied that we could. He told me that before we shipped out to the winter range, I should have a camptender come to the ranch and pick them up. The Home Ranch workers would be done with them in a few days. Ralph said I had better send someone who could drive a young team, as the horses were powerful, three-quarter brothers, and weighed 1,700 pounds each. I arranged for Fred Martinez, an able young camptender, to drive a pair of draft horse "retirees" to the Home Ranch and exchange them for the new young team. Fred was impressed, and said they could be the best on the outfit. He was smiling proudly, holding back on the lines, as he drove away from the barns at the ranch.

Hay stacked loose as was the practice at the Home Ranch.

Will Sorensen had instructed me to always keep seventeen two-horse teams of draft horses for use in operating the sheep outfit. The company always kept a Percheron or Belgian draft stallion at the Home Ranch. Ralph Moss liked to use a Percheron for a few years and then breed the daughters to a Belgian. This system provided big, gentle draft horses. It was easy to select matched teams. Over the years, the company had used two Belgian stallions from the University of Nebraska. Their offspring were docile and easily broken to work. Each time a new team was put into service, it was a great source of satisfaction to cowboys, camptenders, sheepherders, and everyone. The company also raised top-quality saddle horses. One or two Thoroughbred, Quarter Horse, or Morgan stallions were kept at the Home Ranch to sire saddle stock. Brood mares with suckling foals were turned out on the range after the mares were again safely in foal. Cowboys often stopped to admire the yearlings and two-year-olds grazing on the range. About sixty saddle horses were used on the sheep outfit. Cowboys got first choice of the young horses. Sheepherders were given those the cowboys figured were not quite fast enough or agile enough to be a top cow-horse. Sheepherders and camptenders needed horses that were dependable and strong enough to withstand lots of riding. About 10 percent of the horses were retired

Team of matched draft horses at work in the hay fields.

each year because of old age. Cowboys and sheepherders who had used them were sorry to see faithful and dependable old horses go.

Early in September, Will told me that we needed to talk to Jim, the buck-herder, and see how the feed was holding out in the buck pasture. Later that day, Jim told us that it was time to gather the rams and drive them over to the Crane Ranch. He would need a couple of camptenders to ride down there and help, so we made that arrangement. Gathering over eight hundred rams and driving them ten miles worked easier if there was more than one man. The rams were gathered and, along with Jim's camp, moved to the designated area, where there was fresh feed, fenced pastures, and plenty of water. The rams stayed in good condition. In October, they would be shipped to Skull Valley, where they would be fed grain and hay to get them in good condition for the breeding season, which would begin December 15.

Walter Dansie, Will Sorensen, and I decided to start shipping lambs to market about September 12. Will indicated that along about the second week of September, our lambs on the south end would have grown about as much as they were going to on the available feed. We put seven of the herds of ewes and lambs on the higher range, passing them through the corral at the head of Horse Ridge. The whitefaced ewe lambs were saved as replacement ewes, and were left in the herd. All wethers (male lambs that had been castrated) and blackfaced ewe lambs were to go to market. All sale lambs had been given an easily identifiable earmark at docking time, and were separated from the herd by putting the entire herd through the chute. The lambs that were to be sold were separated into a large holding corral, along with all of the older ewes that had been branded with a blue dot after shearing. This separating process took one and one-half days. I operated the dodge gate on the chute and counted the market lambs as they were cut out from each herd. Another fellow counted the old blue-dot ewes. When the operation was completed, we had right at five thousand market lambs and two thousand aged ewes, a total of seven thousand sheep to be driven in one large herd to the shearing corral at Wahsatch.

Will Sorensen wanted to go on this drive, so I suggested that one of our men take a horse and saddle from here at headquarters to the south-end corral. He said that was not necessary, as he had told Joe to have a horse and saddle for him. Four well-mounted men were in control of this giant herd as the sheep were turned out of the corral. Joe had not arrived with a horse for Will, so he started the drive on foot. It was a one-and-a-half-day drive, and they spent one night on the trail. By taking this many sheep off the south-end summer range, the remaining ewes and ewe lambs were combined into three fewer herds. This freed three covered campwagons, with food and beds, to accompany the drive to the shearing corral. We sent baled hay, so that the horses could be tied up during the night.

I met them near the shearing corral the next day just at dusk, and Will Sorensen was still on foot. He was not smiling. Joe was riding a horse, and got to me before Will did. Joe told me that he had informed Will there was not an extra horse on the south end, and that Will was mad and had walked all the way. Will would not use Joe's horse, but Joe had offered several times for Will to ride while Joe walked. As we were driving this herd of some seven thousand sheep into a large, fenced-in pasture for the night, another of the men rode up and also informed me that Will had walked all the way. Will had told them that when he ran the outfit, people were given horses to ride. I walked close to Will and said that I hated to see him walk all that distance from the summer range. He replied that he had blistered feet, and that he was going to soak them in strong saltwater.

I had arranged with Tom Judd to move down to the shearing corral from off the high summer range and open up the cookhouse. He would cook for the crew as the south-end lambs and old ewes were processed through the corrals at Wahsatch, and now he had a huge dinner waiting for the trail crew.

Swift and Company had purchased the lambs, and the next morning their lamb buyer was on hand to separate the market-ready lambs from those that were smaller and needed to have more feed, commonly called "feeders." He looked over the sheep and the corrals. He told me that he would use the first gate and cut the prime and choice lambs into that corral. He went on to explain that he would count the lambs he cut out from the herd, and I should do the same. He said that he used a silver coin, a half dollar, to draw a line across the top board on the chute every time his count reached one hundred. When we were done, we would have an accurate count of each group of lambs. This experienced buyer knew what he was doing. He did not feel the lambs to determine their condition, but separated them according to his visual appraisal as they came single file through the chute toward him.

The lamb buyer wanted to come out even on lambs and railroad cars; that is, he didn't want the last car to have just part of a load. The railroad cars were double-deckers, and each deck held about 125 lambs. He knew he could put 250 lambs in each car. When there were about 1,500 sheep remaining to go through the chute, we stopped the sheep for a few minutes while he and I walked through the remaining sheep. The buyer had to determine if there were only two more carloads of fat lambs, or if there were three. He slightly adjusted his grading and separated enough market-ready lambs to fill three more railroad cars.

With all the lambs separated out of the herd, the old blue-dot ewes went into a different corral. Although they had been mouthed at shearing time, Will Sorensen explained to me that we should mouth them again in order to determine those that had teeth good enough for another winter on our western ranges. Those with teeth that were not good enough would be culled and shipped to an "order buyer" in Omaha, Nebraska, who would again mouth these old ewes and send the "gummers," those with no teeth, to the slaughter plant. The remainder, better ewes designated as "one-year-breeders," were used to fill his purchase orders from Midwest farmers who wanted to run them on their farms for another year. These older ewes would be well fed on those farms and would produce a heavy lamb crop the following spring. When the lambs weighed about a hundred pounds, they would be shipped to a central market, along with their mothers. Midwest farmers liked these older, western-range ewes. The order buyer at Omaha purchased thousands of old ewes each year. Getting these old ewes shipped a thousand miles away was a good thing for the Deseret Live Stock Company. It prevented any cull ewes carrying the company earmark from being marketed locally.

I persuaded Will to sit down and let the younger men examine teeth. We cut out a thousand rejects and drove them to a holding pasture, where they could graze until it was shipping time. The thousand blue-dot ewes with satisfactory teeth were retained for another year. A herder was put in charge of these, and they were to graze a few miles north of the shearing corral. The prime and choice market-ready lambs, commonly called "fats," were driven three miles to the railroad shipping corrals at Wahsatch, where they were weighed and loaded onto Union Pacific railroad cars for shipping to one of Swift's packing plants. The feeders were then weighed and loaded, followed by the reject ewes, all heading east on the same train.

On the high summer range, we began separating the north-end lambs on September 20. It took one and one-half days to put eight herds of about 1,200 ewes each, plus their lambs, through the north corral. Again, all whitefaced ewe lambs were retained for replacements. Over 6,000 wethers and blackfaced ewe lambs were separated out into the large holding corral, and there were 3,300 old blue-dot ewes cut in with the sale lambs. It took four well-mounted men, and two and one-half days, to allow this large herd of 9,300 sheep to graze as they covered the twenty-seven miles to the shearing corral. Three extra camp-wagons accompanied this drive, and there was baled hay for the horses.

While these sheep were being trailed to the shearing corral, Will Sorensen, Kathryn, the kids, and I closed up summer headquarters on the high range and moved back down to our houses at the shearing corral. Mr. Dansie made arrangements with Emma Zabriskie (Mrs. Lute Zabriskie) from Spring City, Utah, to come to the shearing corral and do the cooking for the crew when they arrived with the north-end lambs. There would be a crew of several men eating there until shipping time in early November. Emma had cooked here previously, and we were lucky that she came to help.

The trail herd arrived on the evening of the third day and was put inside a holding pasture for the night. The horses were turned loose into the horse pasture. Emma had prepared an excellent dinner. The next morning the lamb buyer arrived just in time for breakfast. When the sheep were gathered into the corral for the separating process, the buyer told me that we would handle the process the same as we did last time. I told him that we had a higher percentage of fat lambs on the north end because the summer feed stayed green a little longer, and the north-end lambs were heavier. The fat lambs, feeder lambs, and 1,400 reject ewes were shipped on the railroad. The 1,900 blue-dot ewes retained for another year were driven to the herd that remained from the earlier processing.

We had now separated sale lambs and old ewes from fifteen summer herds, and there were five more herds of ewes and lambs to process. These five, in a lower elevation on the summer range, would graze their way to the shearing corral, and all five herds would be worked in one day late in September. When these last five herds were processed, I told Will that we had just over 3,800 blue-dot ewes whose teeth were good enough for us to keep. He advised me to brand all of them with a fresh blue brand, and then take 1,500 of them, add 1,300 replacement-ewe lambs to make a winter herd of 2,800, and send them all to graze to the north toward the little desert, south and east of the Home Ranch. As winter came on, they would be fed hay near the Home Ranch. Will also said I should take the remaining 2,300 older ewes and add 500 replacement-ewe lambs. Then we would have a winter herd of 2,800 to be fed pellets and kept close to the ranch in Skull Valley.

Our young son Bill was four years old on September 27, and Kathryn's birthday was two days later, on the twenty-ninth. We celebrated by getting into the pickup and going to Evanston for ice cream and cake. Modest presents made the birthdays come alive.

Work was proceeding nicely as the hay harvest on the Home Ranch was completed. The thousands of cattle would be well fed during the cold winter. The sheep outfit was preparing to ship to the winter range. In mid-October, I was getting ready to take Kathryn, Bill, and Diane to Evanston to board the train to Manti, where they planned to visit our parents until after the sheep were shipped to Skull Valley in November. I had just put their suitcases in the back of the pickup when up rode Jim, the buckherder. I could see he was in some kind of physical trouble. He gingerly dismounted and was in pain. He tied up his horse as young Bill and I walked over to him. Jim said he had to get to a doctor. His bladder was full of urine and he couldn't get rid of it. He wanted to go to Evanston, so I told him that we were just leaving for there.

I told Kathryn about Jim's predicament. We put the two small children safely into the back of the pickup, and Kathryn, Jim, and I rode in the front. Jim was groaning with each breath, and every little bump in the road made it worse. By the time we covered the three miles to Wahsatch and eleven more to Evanston, beads of sweat were rolling down his face. We pulled up to the clinic, and Jim was on my heels as we entered the door. No receptionist was visible, so I knocked on the door to the doctor's office. Jim grabbed the doorknob and said he was in pain and was going in. The nurse took one startled look, and Jim told her of his predicament. She said for him to come with her, and they went into another room. In less than five minutes, Jim came out with a smile on his face. They fixed him, and he knew what to do if it ever happened again. He felt good; what a relief.

I told him that Kathryn and the kids were going to board the train in a couple of hours, and that we wanted to do a little shopping first. He said that he would also do some shopping, and I should pick him up at Freeman's Restaurant after my family left. Kathryn, Bill, Diane, and I visited a couple of stores and then went to a restaurant for lunch, to have a treat for Diane's third birthday, coming up on October 22, as they would be in Manti then. We enjoyed this time together in Evanston, and I gave them lots of hugs and kisses as they boarded the train going to Salt Lake City and then on to Manti.

Shaping Up for Winter and Shipping to Skull Valley

In October, when the market lambs and cull sheep had been shipped east, Will Sorensen and I were discussing the work still to be done. I told Will we had 41,800 ewes and replacement-ewe lambs as we prepared for the winter. He told me that I should put these in fifteen herds for the winter, about 2,750 to 2,800 in each herd. He liked to put 500 to 600 ewe lambs in each winter herd. I told him we had 7,800 replacement-ewe lambs, and the way he had done it in the past would work again that year. Earlier, I had ordered four cases of red branding paint, and four cases of black, so that we could put fresh brands on for the winter. Each case contained six one-gallon cans with clamp-on lids. The replacement-ewe lambs were too immature to breed this season, having been born in May and early June. Therefore, they could winter with the ewes without danger of being bred by the rams. During the breeding season the following year, they would be eighteen-month-old yearlings, and would breed and have their first lambs at two years of age, the gestation period being about 145 to 150 days.

Will explained that to make up our winter herds, I would have to put all herds through the chute and separate the ewes according to their grades of wool. He indicated there would be three or four herds of "fine" wool, six or seven herds of "half," a couple of herds of "three-eighths," and a herd of "quarter." That way, we could put the right kind of bucks with each herd for breeding. I had a strong background in wool science, having spent several months working in the wool laboratory at Utah State University. As part of my master's degree, I had commercially graded twenty-five thousand fleeces at Charlie Redd's ranch in southeastern Utah. I knew from last spring's shearing that 75 percent of the sheep on the Deseret ranch had a Rambouillet genetic base, from the famous breed originating in France, and produced wool grading either "fine" or the next finest grade, "half-blood." These grades were used in high-quality men's and women's suits. The remaining 25 percent of the ewes had been sired by coarse-wool Romney rams, and consequently produced wool grading "three-eighths-blood" and "quarter-blood," the type of wool woven into bulky sweaters.

This process of visually grading sheep as they came single file through the chute, of separating each animal into one of the four grades, would take several days of hard work. When the first herd was put through the separating chute and the ewes were segregated, they were crowded into the branding chute and given a fresh paint brand for the winter. Each winter herd would have a different brand. Will told me to use black paint brands on our seven south-end herds, and to use red paint on the north-end ones. We could use the same brands with either color paint. Brands commonly used were

$$0, +, T, S, 1, 5, 7$$

Each bunch of sheep had to be counted as they were processed, and then put into their designated winter herd. I counted and kept the tally in a pocket notebook: 2,750 to 2,850 in each herd.

Mornings were cold at this 6,800-foot elevation. A breeze, or more often a stiff wind, across the treeless rolling hills near Wahsatch made working in the corrals "a cold, dusty picnic," according to one herder. The dust irritated the men's eyes, and when they coughed and spat, it was tinged with brown dust. When the men washed up for noon dinner and again before supper, they dug dust out of their ears. At noon the water was cold; however, if the crew was not too large, Emma invited them to wash in warm water at the cookhouse. In the evening, there was time to build a fire in each house or wagon and heat water before our 6:00 p.m. supper.

Emma's huge, tasty breakfasts at 6:30 a.m. got every day started with upbeat optimism. The crew, only a skeleton compared to that at spring shearing, ate at one table in the long dining room. In addition to Will and me, there were one or both riders; Austin Christofferson, who drove the truck and helped at the corrals; Bill Watts, who was the fence crew foreman during the summer; and Tom Judd, the predator-control man who helped wherever needed. At the noon dinner, there were also the herder and the camptender for the herd being worked, plus camptenders who came in for mail and to help in the corral. Emma baked hot rolls for dinner and supper and told me to keep her in plenty of fresh meat, either beef or mutton. We had cured ham and bacon, and the pantry and commissary had all kinds of canned vegetables and fruit. She served a change of roasts, steaks, and chops, along with potatoes and gravy, vegetables, a variety of beans, salads, fruit, and pie. Every day ended with satisfaction in the belly and good-natured banter across the table.

As each herd came into the corral for grading, the majority of sheep were of a similar grade, having been in the same herd since last year's grading. When all herds had been worked, we had four herds of fine, six herds of half-blood, two herds of three-eighths, and one herd of quarter-blood. Thirteen herds were graded according to the wool they produced, and two herds of old blue-dot ewes were put together solely on age. The breeding plan that Will Sorensen and Walter Dansie had used for several years was to breed the Romney rams to the fine-wool ewes to increase the length of the wool fiber, and to breed the Rambouillet rams to the half-, three-eighths-, and quarter-blood ewes in order to make their fleeces denser. This plan would be carried out again this year, when rams were put into the ewe herds December 15 and 16, the beginning of the breeding season.

I wasn't pleased with the coarse-wool sheep, particularly the quarter-bloods, and asked Will if he had considered using Columbia rams instead of Romneys in places where we had so much brush on our ranges. Will did a slow burn and told me that he recommended to Walt Dansie that they use Romneys. I had hit a tender spot, criticizing Will's breeding program. I said that I liked Romneys, but preferred Rambouillets and Columbias for our brushy ranges. Will replied that the Romney crosses don't have much wool on their faces, which helped reduce wool blindness. "Wool blindness" is a condition caused by dense wool growing close to the eyes and blocking the sheep's vision. To prevent wool blindness during the winter and following spring, wool had to be shorn from around the eyes of about 25 percent of the ewes.

Shearing around the eyes was done right after the winter herds were made up. Sophus Christensen, from Bountiful, Utah, was one of the top shearers used for the annual spring shearing, and he came back in the fall to "shear eyes." Each herd was put through the chute at the shearing corral, and those needing wool shorn from around their eyes were separated into the large holding shed, while the main herd went back out onto the range so they could keep grazing. The main shearing shed was not used. Two side-by-side temporary pens held fifty sheep each, and were repeatedly emptied and filled as each lot of sheep had their eye wool shorn. When shearing eyes, it is not necessary to lay the sheep down, as it is in ordinary shearing. The shearer simply walks into the pen of crowded sheep and restrains each one for a few seconds with one hand under the jaw while wielding the power-generated shears in the other hand. A skilled man can shear the eyes of two thousand sheep in a day, and about eleven thousand in our herds needed it. Sophus did the job in six days.

The wool was swept up and stuffed into a bag as each pen was emptied. This shorn eye wool was a short-fiber product, but could be used commercially in some types of felt.

As deer hunting season approached in the fall, it was easy to see why deer on the range should be harvested. They were on all the ridges and in every draw. For several years, this one included, the company had issued five hundred special permits, at no charge, to allow deer hunters to hunt on company property. Most of the range was fenced. In order to control entry into the ranch, we put a man and a campwagon at each of the four entrances. One man camped by the gate at the mouth of Heiner's Canyon and was instructed to allow entry only to those hunters having a permit issued at the company office. Another was stationed at the Scarecrow Gate, an opening in the log fence on the north end of the summer range. A third guard and camp was located at the gate at Wahsatch, and one of the employees at Home Ranch headquarters checked hunters entering from that direction. Hunters were requested to be careful and to make sure it was a deer they sighted on before pulling the trigger. They were also asked to report any sheep that might be lost from a herd. Most hunters bagged a deer.

The eight extra campwagons used only in the summer were stored in the big warehouse at Wahsatch for the winter. Will told me to have the food and bedding taken out of these extra wagons and distributed to some of the winter wagons. When the extra ones were inside the warehouse, the mattresses were removed from the wagons and placed on an elevated metal platform, where mice could not get at them. In preparation for going to the winter range, Will counseled me to order ten new large horse blankets for the draft horses and ten for the saddle horses. A blanket lasted about three winters. I had our two riders check with their camptenders on this before ordering. Each winter outfit had a team of two draft horses for pulling the wagons and two saddle horses for riding. There had to be four good horse blankets with each outfit.

Each winter campwagon, called a "covered wagon" in pioneer days, had at least two, and sometimes three, heavy canvas covers over the top. These covers kept the camp warm and dry. Also, each campwagon had "side skirts," another piece of canvas that circled the lower part of the wagon and reached from the wagon-box, where the top cover ended, to the ground. These skirts were rolled up when the wagon was being moved, but when camp was made, the skirts were let down around the outside of the wheels. This kept the wind from blowing under the camp, and kept the area under the camp warm and dry during winter. The sheep dogs' favorite place at night was under the campwagon. They could get in out of the cold, and if a bitch was going to

Union Pacific depot at Wahsatch.

have puppies during the winter, the camptender fixed a special box under the wagon so the mother and new puppies would be warm and dry.

The crew at the shearing corral awoke one morning to four inches of snow. At breakfast, I said it would be a good thing if the snow hit the winter range, as it provided the necessary moisture for livestock during most of the winter. Sheep eat snow and get by very well that way when water is not available. Will told me that when there was no snow on the winter range both in the fall and again in the spring, we would have to pump water for eight herds. This came from two wells we owned, one on the south end in Skull Valley and one on the north end, on the west side of Cedar Mountain. Each well could water four herds, two per day. The remaining two herds could water at flowing springs. Thus, each herd got pumped water every other day.

The sun came out bright across the rolling hills near Wahsatch, and by late afternoon, most of the snow on the south slopes had melted. On the shaded north slopes, the snow settled a bit, but the freezing nights kept the snow intact. At noon the next day, Walter Dansie drove up in his heavy dark Cadillac and, with a pleasant smile, announced that the storm laid down several inches of snow out west on Cedar Mountain. We would have snow to start with on the winter range.

Shipping to the winter range in the fall of 1946 was going to be a new experience for me. Mr. Dansie and Will Sorensen had been doing it for several years. They assured me that everything would go like clockwork. We decided the best dates for shipping would be November 4, 5, 6, 7, and 8. Will had told me that we should go west with fourteen herds. We would ship three herds each day for four days and just two herds the last day. Mr. Dansie already had it figured that we could leave a herd on the little desert over toward the Home Ranch, and told Ralph Moss that he and the cowboys would have to look after that herd. There was a lot of good winter feed for sheep on our side of the Utah/Wyoming state line, and when the sheep needed hay, the cowboys could feed it to them. Our conference concluded, Mr. Dansie then stopped at the depot at Wahsatch and ordered the necessary railroad cars.

The Union Pacific Railroad responded in plenty of time, and each day provided a special train composed of forty-two double-deck sheep cars, two gondola cars for hauling the wagons, and a single-deck livestock car for the horses. The herder and camptender in charge of each herd were notified well in advance of the day that herd was going to be shipped. Everyone knew the order, or sequence, for proceeding to the railroad. Will advised me that I should allow only the three herds to be shipped the next day to come in close to Wahsatch. The others should stay well back on their range and not push too soon toward Wahsatch. I had cautioned everyone to be on the alert to prevent mixing, and told every herder we wanted a couple of miles between herds when they bedded down for the night.

The day before shipping started, Austin Christofferson went ahead to the winter range to receive the sheep at Timpie, the unloading station on the Union Pacific Railroad at the north end of Skull Valley, about forty miles west of Salt Lake City. He left with a truck, and pulled his housetrailer. Joe Manzaneros was to be responsible for seven herds on the south half of the winter range, and Bill Green for six herds on the north. It would work best if Will Sorensen and I took responsibility for the fourteenth herd, which would be wintered in the foothills north of ranch headquarters in Skull Valley. The south-end sheep were to be shipped first, as their range was the most distant from the unloading point.

Will told me to have Joe Manzaneros go on the train the first day of shipping, so he could help unload the sheep and get the herders headed toward their ranges. Joe assured me that his campwagon would be packed and ready to be towed to the railroad at seven in the morning. He would ride his horse, check on his herds, and then help us load the wagons. On shipping day, each of the three camptenders was to have his team of draft horses pull his two wagons to the wagon-loading ramp at 8:00 a.m. The Union Pacific Railroad had built this loading dock on Deseret Live Stock property, just west of the tracks and north about a quarter mile from the station. The railroad had run a spur of tracks from the main line to the end of the loading dock where, the night before shipping, the U.P. had the two gondola cars for the wagons in place.

Emma had breakfast ready as usual at six thirty. Will rode with me, and we hooked a truck to Joe's wagon and were over to Wahsatch at the wagon-loading dock at seven thirty. Will said that we would have to open the hinged end-gates of the gondola cars and lay the gates flat just inside the ends of each car. Wooden planks were then placed to reach from the end of the dock to the inside of the first gondola. Also, planks were placed to span the distance between the two cars. The wagons would be pushed by hand along the tops of the planks and into the gondola cars, a strenuous job.

The first camptender pulled his two wagons to the bottom of the ramp, stopped his team, and dismounted. He then disconnected the trailwagon from the covered wagon. He picked up the lines, climbed back into the camp-wagon, spoke to his big team, and they trotted up the steep ramp with the single wagon. He pulled almost to the end of the dock, near the gondola car, and stopped the team. I stood at the horses' heads to calm and steady them as the camptender climbed out of the wagon and unhooked the team from the wagon. We unsnapped the cross-lines, so the team could be separated, and hung each line over the hame (the long, narrow curved pieces lying on each side of the heavy leather harness collar). We talked softly to the horses, as they were not used to the hollow sound of their heavy hooves on the elevated wooden loading dock. The camptender and I each led a horse single file back alongside the wagon and down off the ramp.

The camptender then inserted the extension wagon-tongue into the metal brackets on the stub tongue of the trailwagon, hooked up the team, and climbed into the wagon. He said, "Easy, boys," and the big team calmly pulled the trailwagon up behind the campwagon. The horses were then unhitched, unharnessed, and their harness laid on top of the load in the trailwagon. The horses were again led single file alongside the wagon and down off the loading ramp. The camptender led his team a few yards to the fence, tied each to a separate post, and gave them a feeding of hay.

Five of us pushed the first wagon onto the planks and on into the first gondola. One man held the end of the wagon-tongue to guide the wagon in a straight line as it was pushed to the far end of the second gondola, and it took all our strength to push it up over the planks going into the second car. When the wagon was finally in place, we blocked its wheels with wooden two-by-fours so it wouldn't roll when the train was moving. With the first wagon in place, we went back onto the dock and pushed the trailwagon into place right behind the first wagon. While the camptender blocked the wheels, the others walked back onto the dock. A stiff breeze was blowing, but no one seemed to notice nor care. Hats were pulled down tight, and conversation was seldom.

The next camptender had already uncoupled his trailwagon from the camp one and was in the campwagon, holding a taut line on his team as they were poised for the hard pull up the ramp. "We're ready," I called. The man spoke to his horses, they leaned into the load in perfect unison, and were almost at a trot when they came up the ramp. I glanced at Will, who was beaming with satisfaction. Those teams of big horses were beautiful on this steep pull, and just one wagon at a time was a lighter load than they were used to pulling. Again, I stood at their heads at the end of the loading dock and spoke softly to calm them as the camptender dismounted and unhooked. This process of loading one wagon at a time continued. A gondola car had space for four wagons, so when the first four were loaded and securely blocked in the far gondola, the planks spanning the distance between the two cars were removed. The heavy metal car door, hinged to the floor of the car, was lifted to its upright position, the handles were secured, and one car was now loaded.

Finally, the six wagons belonging to the three herds, and the campwagon used by Joe Manzaneros, were loaded. Two of the three camptenders climbed into their campwagons and fixed "dinner on the train." Each of them knew ahead of time that after the wagons were loaded, he would need to fix dinner for his herder, himself, and two or three others. I told Will that they had a deal worked out among the three of them so that just two would have to build a fire and cook dinner. It worked smoothly, and we finished eating lunch at eleven. The three camptenders then hurried out on their saddle horses to herd the sheep while the herders trotted in.

The Union Pacific livestock-shipping corrals were on the east side of the tracks, and about a quarter mile south of the station. Will told me to drive over to the station and ask the agent what time we could cross the tracks with the first herd, making sure we had a half hour without a train whistling through at seventy miles an hour. The first herd crossed the tracks at eleven thirty, and was corralled and ready for loading at noon. Mr. Dansie was there to make an official count of each herd. I gave him a slip of paper and told him that it contained the herders' names and the number of sheep in each herd to be shipped that day.

Max Martinez	2,745
Dan Gallegos	2,755
Fred Cordova	<u>2,750</u>
TOTAL	8,250

Tom Martinez and Salazar on the train at Wahsatch, Utah, heading
for Timpie on the north end of the winter range at Skull
Valley, Tooele County, Utah.

He figured for a moment, and said that he would count 106 sheep to the deck. That would require 13 double-deck cars for each of the three herds, and would use 78 decks, or only 39 cars, for that day. The railroad had brought 42 cars, just in case they were needed. When all the sheep were loaded, Mr. Dansie was pleased. He said he was not used to having counts of each herd come out this close, within two or three of what I gave to him. His tally for the day was 8,245 sheep shipped, and all were loaded before four thirty.

A few days earlier, Will told me to send fourteen to sixteen horses each day. That would be our six draft horses and a few extra saddle horses. On the last day, I was to send an extra team and camp outfit to use when we cut out a few thin sheep from each herd, along with the bucks, at the end of January. At that time we would make up an extra herd of thin ewes, and bring them in close to the ranch and feed them hay. Those three teams of draft horses and ten saddle horses were now in a pen and ready to be loaded. Their single-deck, high-roofed car was pulled into place. Joe and I carefully laid the heavy plank loading-platform in its place to span the distance from the shipping dock to the inside of the railroad car. Will indicated that most of these horses had been loaded before. It was amazing how quietly they walked from the loading alley on into the car. Two of the younger horses inside the car snorted, wheeled around, and came out of the car, but we pushed the tail-enders up the loading alley and blocked the door. We spoke softly to the horses, calmed them, and inserted the two-by-ten-inch wooden brace across the opening of the door. The platform was lifted back onto the loading dock, and the car door was slid shut and bolted. The train now chugged in reverse back up the track to the gondola cars at the wagon-loading ramp.

The herders, camptenders, and Joe had unsaddled, so their horses could be loaded. However, saddles, bridles, men, and dogs were now hauled in the pickup trucks back to the wagon ramp to board the gondolas. We cautioned Joe to keep the herders and camptenders on the train, even though it might stop for a few minutes in Ogden or Salt Lake City. Also, I told the train crew to keep the cars carrying the men and campwagons on the back end of the train, to avoid smoke from the engine. The train would be at the U.P. station at Timpie, at the north end of Skull Valley, at daylight the next morning. We told the men that Austin Christofferson would be there in the morning to help unload. The train crew attached the gondola cars to the rear end of the train, following the horse car, and with waves of the hand and "hasta la vistas," the train headed into the setting sun.

This loading process continued for four more days, and each day the counts turned out very well. On the third day, we sent off the seventh herd, which was going to the south end, and the first two herds, which were going to the west side of Cedar Mountain on the north half of the range. Will Sorensen told me that when we were shipping, it was always a good idea to keep one of the riders around until the last day of shipping. If some herder came in short a bunch of sheep, we could send the rider out to find them. Bill Green, the north rider, stayed at Wahsatch and helped load sheep for five days. Each herder brought his herd to the corral with the full count. On the fifth and final day, Bill Green rode on the train in his campwagon.

The count of sheep on the fifth day turned out to have none missing. The total number of sheep shipped was as follows:

Nov. 4	3 herds:	8,245
Nov. 5	3 herds:	7,980
Nov. 6	3 herds:	8,475
Nov. 7	3 herds:	8,413
Nov. 8	2 herds:	<u>5,727</u>
Shipped west:		38,840
One herd left to winter near the Home Ranch:		<u>2,820</u>
Total ewes:		41,660
Total rams:		<u>864</u>
Total sheep, Nov. 8:		42,524

There were also forty-five saddle horses and thirty draft horses shipped to Skull Valley. A total of thirty-two wagons rode the train west: fifteen covered campwagons, plus their accompanying trailwagons—or supply wagons—for the herds, and a wagon home for each of the two riders. Each of these winter herds had two faithful dogs on the train, along with the men and the wagons. Some had a third dog, a pup growing up. Our dogs were predominately collies, either the black-and-white border collie, or the brown-and-white, slightly larger border collie. Mr. Dansie had purchased a well-bred border collie female, which we bred to a neighbor's registered male border collie. This breeding helped us. Some Australian blood had been introduced and was visible in several of our dogs. Dog mating was usually planned by our herders.

With just two herds sent on the last day, we had enough time to close up the houses at the shearing corral and prepare to drive to the winter range after the train was loaded. Emma rode to Salt Lake City with Mr. Dansie and then would return to her home in Spring City. Will drove his "experienced" black Chevrolet coupe, and Tom Judd his own, even-more-experienced, loaded pickup. Bill Watts, next in line, drove the old green pickup, and I brought up the rear in the black pickup as this not-too-flashy procession left Wahsatch. I felt good about being sheep foreman at the Deseret Live Stock Company. However, I needed Will to stay with me until the whole year's operation was familiar to me. His experience and wisdom in handling this giant outfit was becoming a part of me.

6

SKULL VALLEY

THE DESERET LIVE STOCK COMPANY's winter sheep headquarters was at the main ranch in Skull Valley, located in the old townsite of Iosepa, 20 miles south of the Great Salt Lake. "Skull Valley?" I muttered to myself. Just the name Skull Valley stirs the imagination. What kind of skulls? Ancient buffalo skulls? Skulls of wild mustangs? Or could it be human skulls? I had been told that long ago, Skull Valley was a sacred burial ground for Indians. I conjured up visions of all of these and still wondered.

Skull Valley runs south from the lakeshore and is over forty miles long. It is bounded on the east by the Stansbury mountain range, with peaks over 10,000 feet high, and on the west by Cedar Mountain. Skull Valley is ten miles wide in the north end, where U.S. Highway 40, now I-80, crosses the mouth of the valley between Salt Lake City, Utah, and Wendover, Nevada. The valley narrows to four miles wide at the southern tip of Cedar Mountain, then opens to a vast range further on south. This winter grazing range is mostly semi-desert and varies in elevation from 4,200 feet in Skull Valley to some 8,000 feet on top of Cedar Mountain, all suitable for grazing by cattle, sheep, and wild horses.

Skull Valley was not well known by many people; however, it had achieved some fame, other than for its name. The ill-fated Donner-Reed party of pioneers, heading for California in 1846, crossed the valley near the north end. They took on fresh water at Redlam Spring on the west side of Skull Valley, where Cedar Mountain rises to the west, then traveled in a northwestern direction, and crossed Cedar Mountain going west over Hastings Pass. This was the route recommended to them by Lansford Hastings. This route south of the Great Salt Lake was a costly experience. It was supposed to be a time-saving cutoff, according to Hastings, but he had traveled it only on horseback. The

Donner-Reed party lost much time after they left Fort Bridger (Wyoming), as they had to hack a wagon road through the mountains in order to enter the Great Salt Lake Valley from the east. This road helped Brigham Young and the Mormon pioneers when they came West the following summer, in 1847.

A few years later, in 1863, another frontiersman of considerable distinction, Orrin Porter Rockwell, Brigham Young's right-hand man, established a ranch in the southeast end of Skull Valley, at Sheep Rock Mountain at the west foot of the Stansbury Range, south of Johnson's Pass. However, Porter Rockwell made his home closer to civilization, at the point of the mountain at the extreme south end of the Salt Lake Valley, about twenty miles south of Salt Lake City. The Goshute Indians also called the valley home. The famous and romanticized Pony Express route, carrying mail between Missouri and California, crossed the south end of Skull Valley in order to use the lowest mountain passes. The first permanent white settlers came to the central part of Skull Valley in 1869 and occupied the area that later became Iosepa. This area was an oasis in the desert. Free-flowing, freshwater springs gave life to dozens of large black willow trees, and irrigated hundreds of acres of natural meadowland. This was a logical location for the ranch headquarters of an organization with herds grazing this vast area.

While on the summer range in 1946, in a discussion with Henry D. Moyle, the president of Deseret Live Stock Company, I said, "Iosepa, in Skull Valley, is where a colony of Hawaiian people once lived. How did this come about?" He explained that in the last half of the 1800s, the Church of Jesus Christ of Latter-day Saints, or the Mormons, sent missionaries to Hawaii and other islands in the Pacific. Among those missionaries was Joseph F. Smith, a young man who, in 1854, became a missionary in Hawaii at age fifteen. He was a son of Hyrum Smith, who was a brother of the original Mormon prophet Joseph Smith, founder of the church. This young man, influential in converting a large number of those people to the church, worked diligently with the Hawaiians, and they grew to love him. Their pronunciation of "Joseph" came out "Yosepha" or "Iosepa."

In 1880, this same Joseph F. Smith became second counselor to the church president and held this position until 1901, when he became president. By 1888, there were over two thousand Hawaiians and other South Sea Islanders who had migrated to Utah and were living in Salt Lake City. Many of them asked the church to find a location where they could form a community of their own. Moyle said that the church investigated several properties; finally it purchased the ranch in Skull Valley, because it offered the best possibility for such a settlement. Records at the Utah Historical Society indicate the church paid about $58,000 for a little over five thousand acres of private land. During the summer of 1889, the first party of Hawaiian colonists reached the ranch in Skull Valley. They named their new settlement "Iosepa," in honor of their friend, Joseph F. Smith. Church officials organized the Iosepa Agriculture and Stock Company, a church-owned corporation, and this company employed the Hawaiians at a specified wage. The townsite for Iosepa was surveyed and laid out in square blocks and streets, like most Utah towns. There was a church, of course, plus a schoolhouse, a store, and many homes. A pressurized water system was installed. Shade trees and lawns flourished. In 1915, at the peak of its population, Iosepa was home for a little over two hundred people.

I asked President Moyle when and why these Hawaiians left Iosepa. He told me that when the church built a temple in Hawaii in 1916–19, most of the Hawaiians wanted to return to their native islands, and they did. A few of the younger ones preferred to stay in Utah, and mostly they moved into Salt Lake City. The Hawaiian town of Iosepa lasted for twenty-eight years. There are fifty of these people buried in the cemetery east of the old town. He explained that the church then decided to sell the ranch, including Iosepa. The Iosepa Agriculture and Stock Company was liquidated in 1917. The Iosepa town-site and ranch was sold that year to the Deseret Live Stock Company for $150,000. I enjoyed this discussion with Henry D. Moyle. Here was one able man, a lawyer by profession, with information at his command. Also, I drew on the knowledge of Mr. Dansie and Will Sorensen, who enlightened me on many specifics.

The Delle Ranch, eight miles north and east of Iosepa, had only a few till-able acres, but it was also part of the church-owned property. The church sold it to the Western Pacific Railroad for $25,000. The railroad needed the fresh water gushing from a large spring on this piece of land, located in the foothills of the Stansbury Range. The railroad piped this water seventeen miles northwest to their rail station, known as "Delle," in the north end of Skull Valley. Delle was home to a large railroad section crew. All steam locomotives stopped there to take on a much-needed supply of water as they crossed the wide-open spaces on the western range.

The Deseret Live Stock Company owned another ranch, called the "south ranch," which was on the Skull Valley road, only two miles south of the main ranch at old Iosepa. Both ranches had several hundred acres of farmland, irrigated by water from the Stansbury Range to the east. About nine hundred Hereford cows wintered on the two ranches and received some hay; however, they grazed out in Skull Valley most of the year, on a combination of private and public land. The company's sheep ranged in Skull Valley, and on both the Skull Valley side and the west side of Cedar Mountain, in a 250,000-acre area. About 90 percent of this range was federally owned and under the jurisdiction

Will Sorensen told me to always keep seventeen teams of draft horses on the sheep outfit. It was easy to select matched teams. This photo was taken on the Deseret Live Stock Company's south ranch in Skull Valley.

of the Bureau of Land Management (BLM). The Deseret Live Stock Company had a BLM permit to graze 40,500 sheep on this Utah winter range for a period of six months, from November 1 until April 30. Also, the company owned 10 percent of this range, and could graze over 44,000 sheep if desired. In addition, Skull Valley, Cedar Mountain, and the range west of Cedar were home for hundreds of wild horses. Some were considered to be descendants of the old Spanish mustangs, and some were no doubt horses that had escaped from pioneers, early-day trappers and traders, and Indians. The Skull Valley Indian Reservation covered several square miles in the south-central area of Skull Valley, and the Goshutes there owned many horses.

Soon all fourteen of the company's herds were located on their designated range. Four herds on the north end—on the west side of Cedar Mountain— had used the route over Hastings Pass, where the Donner-Reed pioneers had struggled exactly one hundred years ago, in 1846. Bill Green had told me that the road over the pass was steep. The herds of sheep spread out, slowly grazing over the top. However, the camptender had to unhook his trailwagon while his team pulled the campwagon to the top. Then he had to go back and hook onto the trailwagon to take it up. The trailwagons were loaded with one or two 50-gallon barrels of water, bags of oats picked up at the ranch, bales of hay to feed the horses until they reached their range, and wood for the camp

A camptender, with two teams hooked to a heavy load for a short but steep
climb on the Deseret Live Stock Company's winter range.

stove. It was the camptender's job to keep the camp in wood, do the cooking,
move the camp to fresh feed, and care for the horses. Also, each outfit had a
heavy metal tub used by the camptender to melt snow for camp use, and for
the horses, when they were camped far from water.

We had been on the winter range only a few days when I borrowed Will's
personal auto to drive to Salt Lake City, stored it, then took the bus to Manti.
Kathryn and I bought a 1936 Dodge sedan from my folks, as it was in good
shape, and my parents really didn't need two automobiles. The old Dodge
would provide an occasional trip to town for our family. It was a pleasant
reunion in Manti, with both Kathryn's parents and mine. Kathryn told me it
was good to be here with both our families, but she was anxious to get back to
a house of our own. I told her that our house in Skull Valley was a six-room
bungalow the bishop of Iosepa used to live in. We would enjoy inside plumbing and the bathroom. Will Sorensen lived in a room on the north end of the
big old cookhouse, about forty yards south of ours. I am sure he and Margaret
enjoyed all the winters they spent in what was now our house. Glen Hess was
the ranch foreman; his wife, Edna, cooked for the ranch crew, and they were
real nice people.

With Kathryn and the kids back, and me on the job, Will Sorensen told
me that we needed to build a sheep corral near the mouth of Quincy Canyon,

Our six-room house in Skull Valley. The LDS bishop of old Iosepa lived in this house. It had inside plumbing and a bathroom.

over on the west side of Cedar Mountain. At his direction, I retrieved twenty-five old juniper fence posts that had been discarded in the woodpile intended for burning, and picked up two rolls of old woven wire from the trash pile at the ranch. Will found old boards that were really too short for the separating chute, and we loaded them onto the truck, along with the old posts and wire. When Will and I picked up Bill Green at his camp on the west side of Cedar Mountain, Green was happy to help build a new corral. When the too-small corral was built, Bill Green cussed that it was a waste of time to put up such a contraption, and that if he was going to take the time to dig post holes in this dry, rocky ground, he would like to put in a new stout post, not something that was already half rotten.

In Bill's opinion, we had wasted our time. We should have used good material, and built the corral large enough to hold 2,800 sheep. He said the one we had just built would hold only about 1,500 sheep, and the wire was hardly worth putting up. Will advised him that he could bring in half a herd at a time. Green said that that was the problem; we shouldn't turn 1,000 sheep loose on the range if we could help it. Green thought we were going to build a new corral that would help us, and he was disappointed. Will told him that he was better off with the corral we had just built than he was before, and it hadn't cost a dollar for new material. He and Bill Green were not happy with each other. Green looked at me and exclaimed that as I was supposed to be the boss, he wanted to know what I thought. I told him that I had asked Will to stay on the job and teach me how to run the outfit; however, the next time we built a corral, I would like to use good material so that it would last a long time, and that I would like it big enough to hold a herd of 2,800 sheep.

Will was smarting from this exchange and told us that we young guys had never been through a depression when you had to get by without spending money. In a few days, I quietly discussed this corral situation with Mr. Dansie. He told me that Will Sorensen grew up managing without spending any, or very little, money. I should diplomatically persuade Will that we needed to build well, and to use good material that would last. Mr. Dansie went on to say that we should use labor and material as an investment, and that it was not a good investment to build a corral that would not hold a herd of sheep. He told me that he would leave it up to me to build well in the future. We could afford to spend money if it would help improve the operation and make money.

Back at the ranch, Will counseled me that November was half gone, and that we needed to get fresh beef out to the camps. We usually butchered three cattle every three weeks. Glen Hess took a half carcass for the ranch crew, and we got two and a half carcasses for our men working with sheep. The next day, Glen, Will, Austin, and I hung up three market-ready carcasses, suspending them from the sturdy low branches of a huge black willow tree near the corral north of the beef feedlot. It was too cold for flies. The offal was hauled away to the "pit" and buried. The area was clean. Passers-by could see the hanging carcasses as we let them cool overnight. The following morning, each carcass was divided in two by sawing down the backbone. Will and I prepared the beef to go out to the men. Will said he always cut each half into four wholesale cuts: round, loin, rib, and chuck. Each camp got one-eighth of a carcass. He told me to make a note in my book of who got which cut this time, so that we could give every camp a different piece next time. The camptenders hung the meat in the shade on the north side of their wagons. It was cold at night, and the meat kept in good shape for three weeks. An experienced camptender told me that the men liked steaks and roasts, and that he cooked some tasty stew and good soup.

I took Kathryn, Bill, and Diane for a walk through the big, old horse barn and then on to the beef feedlot, where we counted twenty-eight on feed. We also visited the hog houses, and the kids were excited to see a large white mother sow lying on her side while nine tiny pigs nursed. Kathryn and I casually looked over into another pen and counted twenty-one white, market-ready hogs bedded in clean yellow straw. I told her that those pigs weighed a little over two hundred pounds each, and that it wouldn't be long until we were processing pork. Cold, freezing weather came early in December. Will told me to talk to Glen and arrange a hog-killing day. Two days later we butchered those twenty-one hogs.

With the carcasses hung high, and the offal cleaned away, Glen said that we should divide these as usual. He would take a third for the ranch—that would be seven—leaving us sheepmen to take fourteen. He and a couple of the cowboys hauled theirs to the ranch's meat-cutting room. Will, Austin, Tom Judd, and I hauled ours to the meathouse used by the sheep operation. Each carcass was cut into wholesale cuts. Hams and shoulders were trimmed, then put into several 50-gallon oak barrels. Fat was carefully removed from each carcass, cut into one-inch cubes, and rendered into lard by heating it in tubs on the cookstove in the meathouse. The hot lard was strained into 15-gallon metal drums with clamp-on lids, and stored until needed by the camps later in the spring, summer, and fall. Slabs of bacon were trimmed, rubbed with a commercial salt-sugar cure, and stored on tables.

Will then showed me how to make the salty curing brine to pour over the hams and shoulders in the barrels. We had a special tank that we filled with water, then dissolved enough curing salt in the water to float a medium-sized potato, one about half the size of my fist. We poured the brine into the barrels, and Will told me that we would leave the meat in the barrels for fourteen days, before hanging it in the smokehouse and smoking it. Then Will stepped over to his room to get a fresh cigar. One of the men said that he was glad the old man quit pouring salt in when he did, since in some years Will couldn't find a potato, and put in enough salt to float a rock half the size of his fist.

Fresh cuts of pork—that is, the back, spareribs, and tenderloin—were hung in the meathouse for a couple of days, while the trimmings were ground into sausage. Glen brought over a tub of lean beef to mix with the pork, in order to make the sausage leaner. Sausage was stuffed into commercial casings, so the finished product looked like "store bought," four inches in diameter and sixteen inches long. It took three days to process the pork. Sausage and fresh cuts were taken to each sheep camp. The weather was cold, and the fresh meat would keep for three weeks. Glen told Will and me that he would have another twenty-one hogs ready about the last week in February, and that we should plan on getting them processed while we still had freezing weather. That night at dinner, I told Kathryn that beef would be the main source of meat this winter. However, we would handle another twenty-one hogs the last of February. She answered that she didn't know how other outfits fed their employees, but she knew our cowboys and sheepherders were lucky to have fresh beef and pork, along with everything else they got.

Mr. Dansie drove out the next day and told us that we would have a carload of loose, shelled corn arrive at the railroad siding at Timpie in two days. We then walked through the herd of rams, and Mr. Dansie told us that we had those fellows in good shape. He also told us he had bought a set of fifty big yearling rams, and was having them trucked to the Home Ranch to go into that herd. The next day Will and I, and an extra camptender, began unloading the carload of corn. We used scoop shovels to move the corn from the rail car into the truck bed, which backed up to the car door. There was space for only two men to shovel at the same time, so Will and I began shoveling side by side. Will stopped to catch his breath and told me to hold up a minute. He wanted me to silently count to myself how many shovelfuls I did until he said to stop. Then we could figure out how long this was going to take.

So I shoveled at a normal pace, making sure that each shovelful was all the shovel would hold. I was at 112 shovelfuls when Will said to stop, and asked how many I had done "You've done 112?" he said in almost disbelief. "Well, I've only done 100. You're sure you did 112?" "Of course, I'm sure," I replied. So Will started figuring. Suddenly I realized he was not used to being outworked on a deal such as this. However, Will was sixty-three years old, not the man he used to be, and I, at twenty-six, was in my prime.

We hauled several truckloads of loose corn to the ranch, where a power-driven auger elevated the corn from the truck into the huge old granary. Will said that he was thankful for powered equipment. I couldn't agree more. We sacked the last ten tons of corn remaining in each end of the railroad car into burlap bags that weighed a hundred pounds when filled. These were stored at the ranch, and later would be hauled out to the herd being fed. Handling feed this way was hard work, but I told Kathryn that we were here, we had the time, and—oh, yes—the desire.

Kathryn appreciated pressurized water piped into the house, and she was pleased we had a bathroom and hot and cold running water. She said it would be nice to have electricity, but we were used to going without it. She was happy that Will turned Margaret's gas-powered clothes washer over to her. Will showed Kathryn how to start it: just put a little kerosene in the fire shovel, then light it and poke it under the washer to warm it up before it started. At least that was how he used to start it for Margaret. I told Kathryn that I was glad it sat on the back porch and not inside the house.

7

It Happened During Breeding Season

DECEMBER 15 AND 16 WERE the days when the rams were scheduled to join the ewe herds for the six-week breeding season. The rams in the "buck feedlot" at the ranch were being fed plenty of grass/alfalfa hay, along with some grain, and were in strong breeding condition. They had to be. Some of them would walk thirty miles to the most distant herds. We started them on the fourteenth. Will had previously advised me that it was the responsibility of our two riders to drive the rams out to the herds, so I followed through. On the thirteenth, Joe Manzaneros rode in from his camp, which was twenty miles south at White Rock in Skull Valley, and Bill Green arrived from his camp, west of Cedar Mountain.

Earlier that day, Will, Austin, and I selected sixty rams to go into each of the fourteen herds. About forty Rambouillet or Romney rams went into each herd, to sire whitefaced replacement-ewe lambs, along with twenty Suffolk or Hampshire rams to sire blackfaced market lambs. We branded each ram on the top of the back with a fresh paint brand, identical to the one for the ewe herd he would join. While we men were working with the rams, Kathryn, Bill, and Diane walked over to the corrals to watch. After observing a few minutes, she motioned for me to come over, and then said that years ago I told her Rambouillet sheep had superior fleeces and lived longer than blackfaced breeds. That was correct, but I explained that we needed to have some blackfaced market lambs in each herd. The blackface characteristic is genetically dominant over a whiteface one. Therefore, all lambs sired by Suffolk or Hampshire rams would have a partially black face and a meatier carcass. Buyers liked to see a substantial percentage of blackfaced lambs. We would sell all the blackfaced lambs and keep only whitefaced ewe lambs for replacements.

The next morning, Joe Manzaneros left with 420 rams, heading for the central corral near the landmark White Rock on our southern range in Skull Valley. There they would be fed and held overnight. Joe, with his seven camptenders, had arranged to be at the corral the next morning to help separate the group, and then each man would drive rams to his herd. Later, after Joe left the ranch, Bill Green drove 360 rams west across Skull Valley to the Eight-Mile corral. Next morning, the two camptenders in that area would help separate the rams they were to take. Green would drive the remaining 240 over Cedar Mountain to the corral on the west side, arriving just before dark. On the following morning, the sixteenth, four camptenders would help separate them, and then each would drive his 60 rams to his herd. The camptender with the herd Will and I looked after near the ranch picked up his rams on the fifteenth, as it was only a one-mile drive. I didn't like this method of distributing rams to the ewe herds on the west side of Cedar Mountain. It was an exhausting drive, but we were not equipped to truck the rams. Austin told me that many of the rams would "give out" and have to lie down and rest before they started breeding.

A week before Christmas, Mr. Dansie came to Skull Valley and brought a Christmas check for each employee. This was a pleasant surprise. Each person's check was one-twelfth of what he had earned during the year as an employee of the company, figured to the end of December. Kathryn and I were doubly surprised when my check was for one full month's salary, although I started work in June. We expressed a special "thank you" to Mr. Dansie. I distributed these checks to all the herders and camptenders during the next couple of days. Almost every man immediately put the check into a letter to his wife and gave me the letter to mail so the check would reach home before Christmas. These checks helped build morale and loyalty in the men and made Christmas more enjoyable in Utah, Colorado, and New Mexico.

When I returned from visiting all the camps, Will had just taken the pork out of the brine barrels, washed it, and hung it in the smokehouse for the final curing. He told me that we smoked the meat by building a small fire on the dirt floor of the smokehouse. It didn't have a chimney or windows, and we kept the door shut. The meat hung in the smokehouse for several days.

Isolated Skull Valley was the scene of a decorative and festive "old-fashioned" Christmas. Will Sorensen gave young Bill a new Silver Streak sled, and he gave Diane a new doll buggy. Santa Claus, as usual, came unnoticed in the night, and left enjoyable presents. Edna, over at the cookhouse, and Kathryn had made extra cookies and cakes.

At the end of December, Mr. Dansie wanted to make a personal year-end count of all sheep. He, Will, and I planned to ride horseback for most of three days, but would return to the ranch each night. Austin hauled our horses to the extreme south end, and to the west side of Cedar Mountain. When we returned after dark the second night, I told Kathryn these senior partners of mine were wonderful men. It was a pleasure to ride the ranch with them, but the constant breeze made the days cold, and we were riding hard. Mr. Dansie handled it easier than Will.

The third day was particularly tough. Bill Green had his herds available on the west side of Cedar Mountain, but the horses had to travel at a long-striding trot to cover the miles between the herds. It was almost sundown, and we still had one herd to count. We trotted two miles, Green leading the way, with Mr. Dansie alongside. Will and I followed, side by side. Will, now ashen-faced, was visibly sick but silent. We finished counting herd fourteen just at dark. Austin was there with a pickup to haul all of us to the ranch. Bill Green would handle the weary horses for a couple of days of rest. Will said nothing, got off his horse, climbed into the back of the pickup, stretched out flat, and covered himself with a heavy horse blanket, pulling it completely over his head. I asked Will if we could do something for him. He answered, "No."

Mr. Dansie and I climbed into the seat with Austin, and we proceeded east up Cedar Mountain on the road over Rydalch Pass. We made it safely over the top and down the canyon on the east side. We were on the valley floor and heading northeast to cover the last dozen miles, when suddenly the truck was in the middle of a mud hole. The wheels spun, and the truck was stuck fast in the mud. Austin and I finally got the rear end jacked up high enough to lay brush under the wheels. Mr. Dansie philosophically counseled that this wasn't as bad as it could have been. We cut brush and built thirty feet of brush-covered road in front of the truck. We reached the ranch at ten o'clock. Will went straight to his room. Kathryn had dinner waiting, and Mr. Dansie, Austin, and I made up for lost time.

Mr. Dansie visited with us about the outfit for a few minutes and told Kathryn and me that we were adding some positive things to the sheep operation. He said he had never before had the counts turn out as close as they had since we got there. I assured him that coyotes were less of a problem than in some previous years, and the natural feed was adequate. The snow was about right—enough to provide moisture, but not too much so that it interfered with grazing. I added that Will was showing me how everything should be done. Mr. Dansie concluded by telling Kathryn and me that he wanted to encourage us about our future. He then went to bed, in the room reserved for him in our house.

The next morning, Will was feeling much better. Mr. Dansie, Will, and I talked over the upcoming National Cattlemen's Convention. Glen and Edna

Hess, and Will Sorensen, would accompany Mr. and Mrs. Dansie to this convention, around January 12. Glen asked me to make sure all continued properly with the ranch and the cattle, as he and Edna would be gone for a two-week vacation. Several days later, a little before sundown on a warm sunny day in January 1947, I got out of my pickup truck and was headed for the house when Marie Rasmussen, from our south ranch, drove up. She calmly told me that Bill Green's young bay horse, John, just came into the south ranch with the saddle under his belly. She said that Clarence (her husband) was finishing the evening chores, but that he would go with me to take a look and figure out what to do. I thanked Marie for telling me, but that was not good news. Green could be in trouble.

Bill Green was camped in a central location near four herds of sheep. He usually was alone at night in his camp, on the west side of Cedar Mountain. His camp was forty miles from the ranch by road, and twenty miles as the crow flies. Marie and I talked for a moment about why the horse had come into the south ranch, instead of the main ranch. I thought that Green must have been riding somewhere on his south area, which could be on either the Skull Valley side, or the west side of Cedar Mountain. I told her that I would pick up Clarence in a few minutes. I told Kathryn the situation, and had Tom Judd ride with Clarence and me to see if we could find the horse's tracks and see which direction he came from. That horse was young and had a lot of life, but he was gentle.

It was almost dark as we drove out west of the south ranch, and found horse tracks in the light covering of snow. The horse had come along the seldom-traveled road from the southwest. We followed the tracks for several miles, across Skull Valley to Rydalch Pass on the east foot of Cedar Mountain. The horse had come down Rydalch Pass from somewhere on the west side of Cedar Mountain. It was way after dark, and no trace of Bill Green. We stopped, got out, and ruminated about the situation. I fired my .30–.30 rifle into the air, hoping for a response from Green, but all was silent. We couldn't do much until daylight, so we got back in and drove toward the ranch. We stopped a couple of times and looked back over the country, hoping to see a fire that Green might have lit, but everything was in total darkness.

Before daylight the next morning, Tom and I sacked six bags of oats and put them in the back of the pickup, to give weight and traction for the trip up Rydalch Pass. Just the two of us went out. We had field glasses, a couple of winter horse blankets, and an emergency supply box. We followed the horse tracks up Rydalch Pass to the summit of Cedar Mountain. There, we had a clear view of the wide-open spaces to the west. We used the field glasses as we looked over the country. Way out west about five miles, close to Rattlesnake

Point, we could see the southernmost camp Green looked after. The team was hitched to the trailwagon, and was traveling away from the camp. We followed the horse tracks down the road to the west, where the tracks came onto the road from the direction of the distant camp. By now the trailwagon had stopped, still too far away for us to make out what was going on. I drove the truck off the road in the direction of the camp, picking our way slowly through rocks, brush, and little gullies. In a few minutes, we could see the trailwagon had turned around and was going back to the camp.

As the pickup approached the camp, the team of horses and trailwagon pulled up just ahead at the camp. Bill Green's head was visible above the sides of the wagon-box as he lay in the wagon, covered over with quilts and blankets from the camp. The camptender, Ralph Montoya, climbed into the trailwagon beside Green and spooned some warm coffee into his mouth. Tom and I walked to the wagon and all exchanged "hellos." We could see at a glance that Green had lain out overnight, and was suffering from frostbite. His face was brown as leather. When the coffee-spooning stopped for a moment, he softly asked for a little more coffee. Ralph said that when he walked out about a mile to reach his team this morning, he heard Bill call. He went in that direction and found him in the bottom of a six-foot-deep gully. Ralph harnessed the team, got the wagon, and loaded Bill into it as best he could.

Green was beginning to feel better. He'd had a close brush with death. He started talking and told us that after Ralph and the herder, Jake Segura, helped him count this herd two days ago, he rode up the north slope of the big gully a mile from here. The sun had thawed the surface of the ground, but it was frozen underneath. He explained that his horse slipped, and all four feet shot out from under him, pinning Green's leg between the horse and the ground as the horse landed on its side. Green let go of the reins, as the horse was struggling to get back on his feet. Green knew his thigh was broken.

After I told them our story about the horse coming into the south ranch last night, and about our later having to call off our search because of the dark, Green was feeling much revived. He recounted that he had lain out two nights in the washed-out gully. He got a little fire going, but it didn't make much smoke. He kept busy all through the first night, gathering little pieces of wood and brush to keep the fire going. He couldn't walk, but sat on his rear end, and propelled himself backwards on the ground as he kept moving around for wood. The first night went fast, as he was busy keeping warm. Green could now talk quite well between sips of coffee. He said that the next morning, the sheep came out close to where he was in the deep gully, but they grazed away from the edge of the gully and he never did see the herder. He couldn't pull himself up the steep slope to get out. Green went on that during the second night he just wore out, fell asleep, and the fire went out. When he woke up, it was nearly morning. His fingers were numb with the cold. He could see the camp horses grazing close, so that gave him hope that he would soon see the camptender.

After telling his story, Green ate a small bowl of hot stew. However, we needed to get Bill to the hospital. We gently worked a heavy blanket under him, and carried him to the back of the pickup. We had blankets underneath and over him, and sacks of oats along each side to keep him from rolling. His Stetson was pulled down tight on his head. It was twenty-five miles north, along a vague wagon road, to east-west U.S. Highway 40, and then fifteen miles east to the telephone at Delle. Just after we got onto the highway, the pickup had a flat tire. Tom and I had the wheel jacked up when a Greyhound bus approached from the west. The bus driver could see trouble from far away out in the wide-open spaces of the West, so he slowed down, and then stopped. Tom told me he would finish changing the tire if I wanted to ride that bus to Delle and phone for an ambulance. I told the bus driver about our situation, and he told me to get in; there was a rest stop at Delle.

The passengers asked what happened to that cowboy in the back of the pickup. I related how the horse fell on him, and he lay out on the range for two nights with a broken leg. At Delle, I phoned for an ambulance to come from Tooele. They said it would be there in forty-five minutes. In less than fifteen minutes, Tom Judd pulled up at Delle with a much-revived Bill Green in the back. Green had the blankets pulled up snug under his chin, and his Stetson pulled down tight on his head, with his face now a darker shade of brown. Three days' growth of black whiskers helped him appear rugged, tough, and lean. He surely was. As the passengers came out of the restaurant to board the bus, they crowded around the pickup and wished him well. "Good luck, cowboy," was what most said. I told the men with the ambulance to take Bill to the LDS Hospital in Salt Lake City. Since his leg was badly broken, and his fingers and toes were suffering from frostbite. he would be there for several days. I called Jim Circuit at our company office in Salt Lake City, and let him know the situation; he called the hospital. This accident happened while Mr. Dansie, Will Sorensen, and Glen Hess were all attending the American National Cattlemen's Annual Convention, so Jim Circuit also passed the word on to Mr. Dansie.

Although Glen Hess had asked me to be responsible for the nine hundred cows and the ranch during his absence, the ranch crew handled everything as usual. I watched, but didn't interfere. I asked Austin Christofferson to resume his old responsibility as the north rider, but on a temporary basis. He told me that he would be glad to help out as the rider for a short time. He gave me to understand that he did that job for several years, and he was not going to do it again for very long. Bill Green made a rapid recovery and came out of it in good condition, but it was going to be a long time before he could or should ride a horse. Green said he was going to quit working with sheep for a couple of years. I visited him twice during his convalescence, and we kept in touch.

Mr. Dansie had told me to make arrangements for the care of young Bill and Diane while Kathryn and I accompanied the Dansies to the National Woolgrowers' Convention in San Francisco during the last week in January. We had the children stay in Manti, where they shared time between their two sets of grandparents. On January 25, Mr. and Mrs. Dansie, Kathryn, and I traveled by rail from Salt Lake City to San Francisco. Kathryn said the room-ettes, or compartments, were quite a convenience for traveling on the rail-road. Our roomettes adjoined and opened up to a comfortable little "visiting parlor." The train ride from and back to Salt Lake City was enjoyable, as we passed across the broad expanses and far horizons of Nevada and climbed up over the snow-covered Sierra Nevada range.

All during the convention, the California woolgrowers had special favors of nuts, olives, oranges, and other delectables sent to each room. This super hospitality made us appreciate the good life in California. It was a convention to remember, with many inspirational and happy occasions.

One free evening, Kathryn and I went to a fancy San Francisco restaurant. We asked the gentlemanly waiter for a dinner suggestion, and he proposed "Chicken-au-Jerusalem." We didn't know what it was, but said it sounded good. A chef prepared it in a flaming dish next to our table; it was most delicious. The next day, Mrs. Dansie said she wanted to take Kathryn and me to lunch at Fisherman's Wharf. We gladly accepted. Joining us for lunch was Mrs. Wick Stevens, a longtime friend of Mrs. Dansie. The Dansies and Stevenses had been friends in Salt Lake City, though the Stevenses now lived in San Francisco.

Mrs. Stevens told us about an occasion when her husband worked in a bank in Salt Lake City. A well-dressed, very proper gentleman walked into the bank, wearing a high silk hat. He was from Boston and came up to Mr. Stevens and introduced himself. Then he pulled out a document and said, "I have here a mortgage on 1,000 'eewees.' Pray tell me, what might they be, and where might I find them?" Mr. Stevens explained that a "ewe", pronounced "u," was a female sheep. Then he read the document and advised the gentleman about how to proceed.

My specific convention responsibility was to serve on the Lamb Marketing Committee. My name was listed on the program, and I was an attentive lis-tener, who could discuss both problems and opportunities. Mr. Dansie

Diane and "Bill" Dean Frischknecht on our front lawn in Skull Valley, spring 1948. West of the north-south highway is the huge and historic "drive-through" granary once featured on the radio as one of the historic ranch buildings of the Old West, as told by "Death Valley Scotty."

introduced me to several livestock producers, both from Utah and from some other states. This was a growth experience for Kathryn and me. We were proud to be introduced as part of the Deseret Live Stock Company. And living in a fancy San Francisco hotel was a great change from the rugged life on the far-out ranges of Utah.

At the end of January, Austin Christofferson, who replaced the injured Bill Green, separated the rams and about one hundred thin ewes from each of Bill's four herds grazing on the west side of Cedar Mountain, and drove them east over the mountain to the corral at Eight-Mile. He did the same with the two herds that were in that area and drove all of them to the ranch in Skull Valley. He gave me the count of each herd. Austin could do that job very well, but he just didn't want to do it anymore. Joe Manzaneros, on the south end, cut out the rams and thin ewes from his three herds on the west side and four herds on the east side, and counted each herd at the same time. He drove his sheep on in, arriving after Austin. Joe and Austin were able and dependable men.

We separated these sheep and put the rams back into their corrals, where they could get a full feed of good alfalfa/grass hay, but no grain. We put the thin ewes into what we called a "scad" herd, a word sheepmen use for thin or maybe even cull sheep. I put a herder and camptender with that herd, and they used the extra wagons we had shipped the previous November. This herd grazed east of the main ranch, toward the foot of the Stansbury Range on the east side of Skull Valley. Each day the camptender hauled a hayrack wagon-load of bright green alfalfa hay from the ranch. The hay was fed to the sheep each morning, so they could get some feed into their bellies before going out to graze on the range. The sheep responded very well to the extra feed, and soon it was obvious that they were gaining weight.

It was convenient to have the old Dodge for personal travel. Kathryn drove alone to Manti and picked up Bill and Diane. It was a happy time when our family was reunited in Skull Valley.

Old-Time Range Strategy

GLEN HESS DID LOTS OF riding, knew the country where everybody's cattle grazed, and knew where most of the hundreds of wild horses "hung out." He always conferred ahead of time on things we did together. We butchered another twenty-one hogs on February 20, while the nights were still freezing. This process was a repeat of the one in December.

That winter our family became acquainted with some enjoyable people. Jess and Leona Charles owned and operated a restaurant, motel, and service station on U.S. Highway 40, a few miles west of Grantsville. Whenever we left Skull Valley and went to town, we stopped to visit this family with its five daughters and one son, some about the ages of Bill and Diane. I was sitting alone at the Charleses' lunch counter when a fellow from Grantsville came up and introduced himself. He told me that he understood I was the new sheep foreman for the Deseret Live Stock Company. I said that I was "learning the ropes" from Will Sorensen. The man wished me good luck and said that I'd need it running forty thousand sheep, with Mother Nature as my best friend and worst enemy. I would have to manage to get through summer droughts, he continued, followed by too much snow in winter. How true. A mile west of the Charles place was the Dolomite lime plant, where Bill and Gene Miller lived. They had three daughters and one son, just a little older than Bill and Diane. The Charles and Miller families invited us to several family dinners. We rushed the season and enjoyed a Dutch-oven barbecue picnic in February, on a sunny southern slope in Skull Valley.

In late March, Mr. Dansie came to Skull Valley to see how things were going and to have a conference with me. He told me that Will Sorensen was elected a director of the company at the annual stockholders' meeting, held on March 10 in Salt Lake City at the Newhouse Hotel. Will had not told me. Mr.

Dansie said that since Will was going to be taking a less active role, they didn't want him to lose interest in the company, so Will was proposed as a director, and the stockholders elected him. He'd been with the company a long time, and this was a nice recognition for him. I agreed that as a director, Will could give a first-hand account of range and sheep conditions and could offer valid opinions at meetings.

The officers and directors of Deseret Live Stock Company were as follows:

Henry D. Moyle, President
James D. Moyle, Vice President
Walter Dansie, General Manager, Secretary, and Treasurer
Ralph Moss, James Moss, Edward O. Muir, and
    William H. Sorensen, Directors

The book value of the company's stock, as of March 10, 1947, was $18.43 per share, with a net worth of $1,300,000.

That evening, when I told Kathryn about Will now being on the board, we mutually knew that our work and total family conduct would have to be on the "outstanding" side of good. The next day, Will accompanied me as we drove around the north end of Cedar Mountain to see the four northern herds on the mountain's west side. I congratulated Will on being elected a director of the Deseret Live Stock Company and added that it would be helpful to have him on the board. He thanked me and said that although he had very little schooling, less than four years in grade school, he had put in a lot of years here and acquired a little stock in the company. Some time ago, he had told Henry Moyle that if a small block of stock came up for sale, he would like to buy it. He was thinking of maybe one hundred shares, but two hundred shares became available, and Henry handled the purchase. Will thought this was a good investment, since he and Margaret never had any children, and now she was gone. He figured putting some money into the company was a smart thing to do.

He noted that he was glad to be a director, especially since he once got fired from his job with the company by Bill Moss, who was its general manager then. Will had been in charge of day-to-day management of the sheep for several years. There were some bad years during the Depression in the early 1930s, with low prices and high death losses. Will had dedicated his life to running those sheep. He gave it his best but was fired. Will felt bad, terrible. Moss thought someone else could do it better. Will said he was gone for about four years. Walter Dansie became general manager in 1933. He'd been in that position a while when he asked Will to come back as sheep foreman. So now Will felt good about being a director. He said the worst part about being a hired man was that you might be doing the best job possible, but if the top brass didn't think so, you could end up somewhere else.

Part of my job was to listen to what Will told me about how to manage the sheep. I planned on doing it right, even making some needed improvements. Someday, if I did a top job, I could be the top man—general manager of the Deseret Live Stock Company. That was my innermost goal, not to be mentioned publicly, but a goal Kathryn and I could work toward. Mr. Dansie had told me to become familiar with the cattle operation up at the Home Ranch, as well as the cattle setup in Skull Valley. Will and I concluded our visit to the herds on the west side of Cedar Mountain, and were returning to Skull Valley by way of Rydalch Pass, near the center of our range. We stopped on top of the mountain and looked east across the valley. Great Salt Lake was easily visible thirty miles to the north, and Will pointed to landmarks thirty miles south. He said that you could always tell when it was spring in this country. We could see separate little storms going on across the valley, counting seven thunderheads of clouds causing small squalls of rain, a sure sign of spring.

As we arrived back at the ranch, I reflected on this enjoyable day. The ranch at old Iosepa had a huge, beautiful yard, with one large, secure fence encircling our house, the cookhouse, and bunkhouse. It provided a safe place for the kids to play. The lawns were greening, the shrubs and flowers budding. Kathryn had raked and sprinkled the lawns and had kept the hose running to the row of tall trees to the west.

During late March, the Deseret Live Stock Company put a third man with each herd. This assured adequate help from then through lambing. Also, we needed herders for the extra herds of ewes and lambs during the summer. One day Mr. Dansie showed me a telegram from a man in New Mexico who had worked for the company previously and wanted to come back. The telegram was addressed to the "Dantasy and Jimmy Sheep Company, Salt Lake." Mr. Dansie said that Western Union did a remarkable job getting such messages to the right place. Someone there must have known Walter Dansie and Jim Circuit were in the office at the Deseret Live Stock Company. The company sent that herder an advance for a bus ticket, so he would be here at some point.

As the weather warmed up and snow on Cedar Mountain receded up the slopes, Will and I prepared to start pumping water from the two wells out on the range. The heavy mechanical pump and the gasoline-powered engine from each well were stored in a shed at the ranch. Will told me that we had to haul the pumps and engines in here every spring when we were leaving the desert. It was a hard job to haul them back out and install them again, but it needed to be done, because the whole works had been stolen when everything was left out on the range one summer.

The weather warmed up in April, with plenty of fresh green feed over the ranges. In the moist areas of Skull Valley, hordes of tiny gnats hovered about. Some of the men called them "No-see-ums." I told Kathryn and the kids that when we saw a herder with his large handkerchief, or bandana, wrapped around his face and neck, we would know it was for protection against swarms of gnats. We had a few mosquitoes, but they were no bother compared to the gnats. One nice thing about it—when the wind blew, the gnats disappeared.

Several times during the winter and spring, when driving through the Goshute Indian Reservation to the south, I noticed a golden eagle perched on a fence post fifty yards off the road. The huge bird was almost a fixture near that location. At the supper table in mid-April, I told the family that when driving through the reservation that day, I saw the golden eagle hanging dead from the fence. Upon examination, I saw that it had been killed by gunshot. Our company policy was against the senseless killing of wildlife. However, when it came to sheep-killing coyotes, we had a different policy. Too many times, coyotes had killed from one to over a dozen sheep, and not eaten a single carcass. This was wanton killing, killing for fun. Will had counseled me that coyotes ate lots of rabbits and rodents, so we didn't want them eliminated, just kept under control. We had Tom Judd employed as a private trapper, and we cooperated closely with government control officers assigned to our areas by Department of Fish and Wildlife officials.

When the snow ran out, the pumps were operating. Bill Watts manned the pump on the north end, west of Cedar Mountain, and lived in a one-room cabin at the well. I visited every few days. One day the two of us drove west a few miles in my pickup, inspecting forage and range conditions. Sheep were being watered every other day and grazed far away from the well on dry days. Grass was lush. Watts asked if I had ever seen the Livingston Pond. I had not, but Will Sorensen told me that at one time Joe Livingston ran a herd of sheep in this country during the winter and spring. Joe was from my hometown, Manti, but years ago went to Craig, Colorado, and became a large sheep operator. I met him on one of his visits to Manti, along about 1937 or 1938. Watts directed me to cut to the left a little and take a look at a pond Joe built. It was well situated to catch water from melting snow during spring runoff. We arrived at a shallow pond thirty feet wide and one hundred feet long. Over the years it had silted in, but even now it stored water six inches deep. Watts explained that it was a good pond when he first saw it. Joe used a horse-drawn scraper to move several hundred yards of earth, quite a job for that kind of equipment. It was built before any wells were drilled out here. Using the pond, Joe could water his sheep for three or four weeks after the snow melted.

I said that everything Will told me about Joe Livingston indicated that he was one shrewd operator and that you had to get up early in the morning to deal with Joe. Watts chuckled and said, "Let me tell you a story." Before the BLM, when this range was still open country and there were no designated permit areas for each herd of sheep, Joe had one herd west of Cedar Mountain. The Deseret Live Stock Company had several herds on the west side as well and wanted to run Joe and his sheep out of the country. One day, four herds of company sheep moved in close to Joe's. The strategy of the company foreman was to have one herd of company sheep mix with Joe's sheep the next day. Then they would have to take the mixed sheep to the corral for separating. As soon as Joe's sheep came out of the corral, a different company herd was to be driven into Joe's, and they would again have to go to the corral to be separated. This process was to be repeated each day, using the four company herds to change off mixing with Joe's herds and going to the corral. It was planned so that Joe would have go to the corral every day, until he got tired of it all and pulled out of the country.

Watts went on to say that Joe was out here with his sheep and could see the tough predicament facing him. But he wouldn't pull out. Early the next morning, the first herd of company sheep was being grazed and pushed toward Joe's sheep. Joe had already told his herder and camptender about the strategy he could see was going to be used against him. But he had brains and guts, and he didn't scare too easily. Watts said that Joe cut off one-quarter of his herd of sheep and pushed them right into the approaching company herd. He then cut off another quarter and pushed them into the next closest company herd. Joe just divided his one herd of sheep into four bunches, and pushed a bunch into all four company herds that were surrounding him. The company foreman then had to put all four of his herds through the corral. When they finally got the fourth herd separated two days later, Joe told them that if they wanted to start this all over again the next day, they could have all of his sheep pushed in with theirs again. That ended the mixing. Watts said that what Joe Livingston did took courage.

I told Watts that Joe moved to the west slope of the Rockies in Colorado and became the sole owner and operator of ten thousand sheep. Some of his nephews from Manti moved to Craig, Colorado, to go into the sheep business with him. Watts reasoned that Joe could see there wasn't much chance of him expanding around here, but that he was the kind of guy who would make it big if he had half a chance. He admired Joe, and said Joe was the underdog in a fight with a giant. Watts told me that he always cheered for the underdog, even if the giant was the Deseret Live Stock Company.

Back at the well, we watched a herd coming in for water. We were pleased to see that the herder rode in front of the sheep, holding the leaders in check as they grazed toward the well. This prevented them from running the last two hundred yards, as they knew they were heading for water.

A few days later, Will counseled me that we needed to look over the old horses in the corral at the south ranch. We identified seven that should be shipped to a packing plant as their final destination. Marie Rasmussen came out of the house to see what horses we were going to let go. She told me that last year she saved Old Bally when the old horses were being shipped. He was such a magnificent horse when Clarence and Austin were the two riders on the sheep outfit that she couldn't stand to see him sold. She had ridden on him lots of times; he was perfectly gentle and would make a great horse for our little kids. I asked Will what he thought, and he said for me to decide. I told them that I was in favor of keeping him for the kids, so Old Bally was saved again. That proved to be an excellent decision. At the end of April, he was shipped back to Wahsatch with the other horses and kept at the horse pasture at the shearing corral and then went to the summer range.

The sheep had to be off the winter range by April 30. Also, they needed to be back east on their lambing grounds near Wahsatch before lambing started on May 10. Mr. Dansie, Will, and I agreed that shipping them east by rail should begin on April 26, and go for five days. A few days ahead of that shipping date, Will explained that I needed to have Austin haul all the cured pork and the 15-gallon drums of lard to the commissary building at the shearing corral. Also, I should have a large grub order made out for him to pick up in Salt Lake City and take with him. He could then come back here and help out a few days before we started shipping. Austin handled it as outlined.

I notified all the north-end herders about the shipping schedule, and had Joe Manzaneros notify his south herders. Back at the ranch, Austin sacked fifteen 100-pound bags of potatoes from the spud cellar, then hooked on his housetrailer and drove to Wahsatch to receive the sheep and get them headed to the designated lambing areas. Will advised me that we should cut out the yearling ewes, so they were not included in the lambing herds. We could do that at the corral he built a few years ago at Muskrat Spring, between here and the railroad. We would want about 2,500 lambing ewes in each herd. I told Will that a few days ago I had walked around the holding pasture there next to the corral. It needed repair in only one small area, so I made the repair, and now it was sheep tight. It would hold the yearlings.

A herd of sheep going into the shipping corrals at Timpie, on the north edge
of the winter range. The sheep would go to Wahsatch, on the spring range.

The previous fall, we kept 7,800 replacement-ewe lambs. We left 1,000
in the herd up at the Home Ranch and sent 6,800 out here. After the winter's
loss, we should have had about 6,500 to ship. Will counseled me that he and I
could separate out the yearlings from each herd the day before that herd got
shipped. We would eliminate two herds by cutting up the scad herd and one
other herd, to make 2,500 sheep in each lambing herd. That way we would
have only two herders and camp outfits to handle the yearlings when we got
them separated.

That evening, I told Kathryn that Will had been through this operation so
many times that he knew automatically how to manage to have enough herders,
camptenders, and complete outfits to handle each herd when some herds were
cut up and new ones created. I then explained how the separating would be
done at the Muskrat Spring corral. Kathryn reasoned that it was a good thing
Will had agreed to stay for at least a year and said that he was a considerate man
and a gentleman. She had never heard him swear, and he was good with our
little kids. That was all true, and he had a keen sense of decent humor.

The previous fall's replacement-ewe lambs had been wintered and were
now referred to as yearlings. They were too young to lamb and should not be
in the "dropherds," a term used for ewes going to lamb, that is, dropping their
lambs. Shipping from the desert went smoothly. I gave the counts of each

Covered wagons and trailwagons heading for some gondola railroad cars, for shipping back to the Wahsatch, Utah, spring range.

herd to Mr. Dansie, who figured how many ewes should go into each deck of the double-deck railroad cars, and counted each herd one deck at a time. Two gondola cars of wagons and one carload of horses were shipped each day. On the third day of shipping, we sent out two herds of lambing ewes and a large herd of 3,700 yearlings. This cleaned most of the yearlings out of the holding pasture. On the fifth day, April 30, the last herds were loaded, which included a herd of 2,800 yearlings.

Kathryn, Bill, Diane, and I loaded clothes and bedding into the old Dodge, and followed Will in his Chevy coupe as we said goodbye to Skull Valley and headed east to spring headquarters at the shearing corral near Wahsatch.

9
LAMBING

WHEN OUR FAMILY ARRIVED AT Wahsatch, the last herd had been unloaded off the railroad, and the sheep were grazing toward their lambing range. It was noticeably cooler here, at a 6,800-foot elevation, than down in Skull Valley. It was exhilarating, driving over the rolling hills for the three miles from the railroad station to the shearing corral. It was May, and we'd been gone half a year. It seemed as if we'd been gone only a week, except now it even smelled like spring. The grass was up six inches. The creeks were overflowing, the reservoirs full. It was amazing to see all the ducks, which were on every pond and along the creeks.

I needed to see where the herds were situated for lambing, and how they were going to be organized into summer herds. It was a special moment, looking across the open country near Wahsatch. I could see sheep in the distance to the north, west, and south, since there were no trees for several miles. I reasoned that if we got a snowstorm during lambing, there would be no protection for newborn lambs in some of our herds.

We opened up our two-room house. I got a fire going in the cookstove, and carried in a couple of buckets of water from the well. Emma Zabriskie had arrived earlier in the day. She had the cookhouse warm and food cooking. Kathryn, Bill, and Diane joined me to say hello to Emma, then we went to work on our home. Airing out the houses, sweeping, and scrubbing were big jobs after everything had been closed up for six months. When it was clean and the beds made, food was carried over from the commissary. Will had helped me make out a huge grub order for headquarters, as we would have a big crew before we started docking lambs in a couple of weeks. This way we could help ourselves to whatever we wanted, since there was plenty.

"Bill" Dean Frischknecht in the saddle, with Diane behind, riding Old Bally at Skull Valley, November 1947.

On my first trip to Evanston, I went into Ballinger and Eastman's store and purchased an exceedingly well-made "Tex-Tan" kids' western saddle, as well as a new bridle to go on Old Bally. Young Bill and Diane were thrilled. I showed them that the two bridle reins were fastened together, just in front of the saddle horn. I explained how to hold the reins in the left hand, so the right hand is free, and said they should lightly "neck rein." To go to the left, they just had to gently pull the combined reins to the left on Old Bally's neck, and do the opposite to go to the right. To stop, they should pull gently on the reins and say "whoa." Old Bally knew much more about it than they did, and he responded to their light touch. Every day they spent hours riding double and taking turns sitting in the saddle. They were allowed to ride to the large lake one-half mile south of the house and could ride one mile north to Romance Rock. However, they had to check back in with Kathryn every little while.

Romance Rock was flat topped, of pale, orange-pink sandstone, five feet high and twenty feet in diameter. It was a good place for a young couple to visit. They could climb atop the rock and gaze at the moon and stars, away from the crowd at the shearing corral. It was even credited with having helped two recent romances blossom into marriage. For Bill and Diane on Old Bally, it marked the outer perimeter of their territory while riding the range alone at ages four and three.

The mixed herd of old ewes and ewe lambs, which had wintered near the Home Ranch, were grazed to the shearing corral for separating. The old ewes were added to the one smaller herd from Skull Valley, to make 13 herds of lambing ewes. The yearlings were added to the herd of 2,800 from Skull Valley, now making two herds of about 3,700 each. Lambing season officially began May 10; however, a sprinkling of lambs were born on May 7, 8, and 9. Shipping the sheep off the winter range the last five days of April gave us adequate time to get every herd to its designated lambing area.

Lambing on the range requires close herding. We had enough range to give each herd a large area, with plenty of feed and water. McKay Creek, five miles north of the shearing corral, was an ideal lambing ground. It was a gently sloping canyon, with a wide bottom and a stream running through it, and there were trees and brush for protection in a snowstorm. Will explained to me how the ewes in each herd were herded into a rather confined area on one side of the stream during the forenoon. About noon, the ewes that lambed that morning were left standing with their newborns, and the main herd of "droppers" was moved across the stream and slightly upstream for the afternoon's grazing. The ewes that lambed that forenoon gradually worked together into a little flock, and their lambs were referred to as the "morning drop," which could be fifty head. In the evening, the drop herd was moved back across the stream, leaving behind those ewes with lambs born that afternoon. This little flock with new lambs would then gradually work together, and stay in a separate bunch. At daylight the next morning, the drop herd was quietly moved upstream and across, and the "night's drop" were left behind.

This system was followed each day, so that during a 24-hour period, each herd left behind three little bunches of ewes with their new lambs. The stream helped keep those bunches separated for several days. The lambs grew, and they and their mothers became accustomed to each other. Will said that in each herd, during a 24-hour-period, about 150 ewes would have a lamb. The camptender could ride around and look over the three little bunches, but it was best if he didn't bother them. After 1,200 ewes had lambed, the herder and his camp would stay with the 1,200, and that would be his summer herd. He then gradually let the ewes work into two bunches of 600 each. That was how we wanted them for docking. The two bunches of sheep would again gradually work together into one herd after docking.

The docking crew at work, 1946.

Will counseled me to have the extra campwagons pulled over here from the warehouse and parked close to the blacksmith shop, so they were handy for repairing. Then we could supply them with food and bedding. He went on to explain that when two neighboring herds had each lambed out 1,200 ewes, the 1,300 ewes that had not lambed were joined to start a new herd of about 2,600. That's when a campwagon had to be taken out for the newly created herd of droppers. When the "second round" had lambed out 1,200 ewes, those that hadn't lambed were joined into new herds. Finally, lambing would be over, and we would end up with ewes that had lost their lambs or weren't going to lamb. That herd was the "residue." As lambing progressed, the new herds kept slowly working upstream.

Arrangements had been made for eight husky high-school boys to help us from mid-May until shearing finished in early July. They arrived in plenty of time to help build a fence, which doubled the size of the buck pasture. That year, six of those young men would probably later work on the summer fencing crew in the high country.

Docking started the last week in May. The term "docking" means to dock, or cut off, the long tail. It was a necessary measure with sheep, because it helped prevent manure from getting caught and accumulating under the tail. Also, having the long tail removed facilitated breeding when the ewes became mature. At docking time, the male lambs are made into wethers, in order to prevent indiscriminate breeding by them before they are marketed in the fall.

Will, Austin, and I were preparing for the first day of docking. We had checked the portable docking corral, and had three rolls of woven wire loaded onto the truck, along with thirty steel posts. Will said that we should do the closest herd first; that way, if we forgot anything, we could come back and get it. I told them that the closest was Manuel Silva's herd, a couple of miles north, so we loaded two gallons of red paint, enough for his herd, and the small branding iron for lambs, getting everything ready to start the next morning.

We had Emma ring the first bell in the morning at four thirty, so we could eat at five. I told the men and Emma that we'd be back for dinner at noon. Will told me it worked best to have ten or more men at docking. We'd set up the corral twice in the forenoon and do the two bunches in one summer herd. Then we'd repeat the same process in the afternoon. That way we got two complete summer herds done each day. As long as we were within ten miles of the shearing corral, we'd come back there for dinner at noon. If we were way out on the range, we'd have to take some food to help the camptender feed us.

After selecting a site on fairly level ground, Will indicated where the corral should be set, and the direction the two wire wings were to take from the opening, or corral gate. These two rolls of wire were laid out to form an "open V" extending out from the entrance to the corral. The wings helped to control the sheep as they were driven to and guided into the corral. As they entered the corral, one wing was removed from the posts and used as a gate to enclose the sheep from behind. A small holding pen for lambs was set up just inside the wire gate. The pen, ten feet long and six feet wide, was made of three wooden panels. The ten-foot panel served as the back side of the pen, and the wire corral gate served as the front side. A wooden plank ten feet long and ten inches wide was placed on top of the front of the pen and served as the working table while the lambs were docked and branded. Four men picked the lambs up out in the corral and carried them to this catch pen.

Three fellows in the catch pen caught the lambs and held them for docking. Each man would grab a lamb, hold it by all four legs, and set the lamb on its rump on the docking plank, with its tail toward Austin or me, the two dockers. Male lambs were castrated surgically and efficiently, without rough handling, using a hand-held, scissorlike tool specifically made for that purpose. The bottom end of the scrotum was cut off, exposing the testicles, which were gently removed one at a time. The next job was to dock the lambs' tails, using the same instrument. The replacement-ewe lambs' tails were left ¾ to 1 inch long. Market lambs' tails were left slightly longer, 1½ to 2 inches in length.

Next, the lamb was presented headfirst to Will, who did the earmarking. We three and the brander stood on the other side of the plank, on the outside of the corral. A sharp pocketknife was used to apply the Deseret Live

Stock Company's ownership earmark. This mark was registered with the State Department of Agriculture. Replacement-ewe lambs were given the earmark shown below.

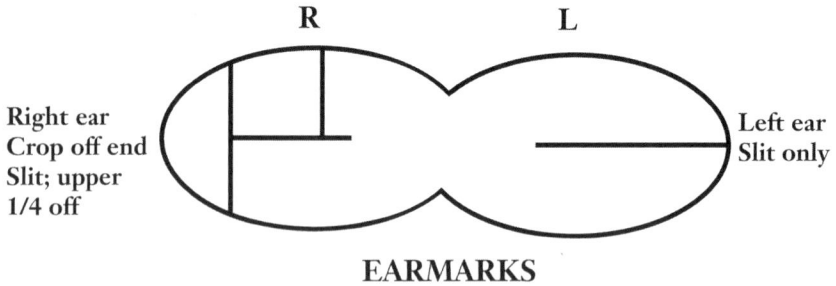

R   L

Right ear
Crop off end
Slit; upper
1/4 off

Left ear
Slit only

EARMARKS

Market lambs received a simple "upper slope" mark on the right ear only.

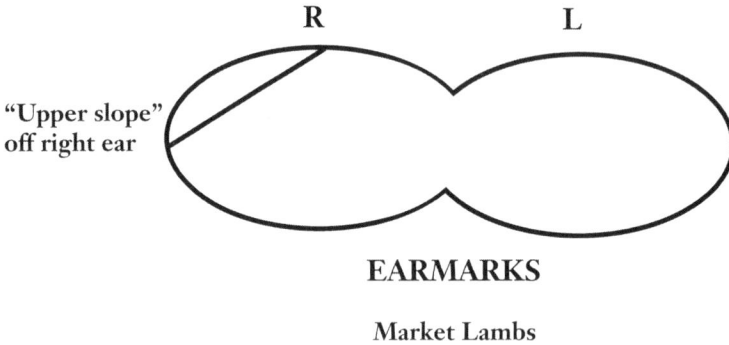

R   L

"Upper slope"
off right ear

EARMARKS

Market Lambs

Putting a paint brand on the back of the lamb was done last. The holder then leaned down and set the lamb gently on the grass outside the corral. One fast earmarker, such as Will, could keep up with two or three dockers.

After each corralful was docked, I counted the ewes coming out of the corral gate, while two fellows counted the tails into piles of one hundred each. It was a real satisfaction if six hundred ewes brought in six hundred or more lambs. Lambing in the high country in May that year had subjected the new-born lambs to one cold storm, and some lambs didn't survive the freezing temperatures. As soon as the ewes were counted, the crew took down the corral, rolled up the wire, uprooted the iron posts, loaded everything onto the truck, and proceeded to the next location. After the first couple of times the crew set up the corral, each person took responsibility for particular procedures. I told the boys that we were beginning to function like a professional circus crew, setting up, taking down, and moving.

Thirteen summer herds were docked on the first go-round, and it took seven days. By then, the lambs in the next age category were old enough. This

The docking crew at noon dinner out on the range, 1946.
Will Sorensen, front right; Dolan Downard, front left.

process continued, and all except the youngest herd of lambs were docked before shearing started June 12. When we docked in Heiner's Canyon, on the far south end, and down on Salaratus Creek, on the north, Emma sent large kettles of stew, homemade bread, and extra eating utensils to help camptenders prepare the noon meal. The camptenders were most cooperative in preparing extra food for several men. Western hospitality operated at its best at our camps.

I drove north to the "little desert," southeast of the Home Ranch, and counted the two yearling herds on June 1. The counts came out very well, and all were accounted for. That evening I told Kathryn about counting the yearlings of a new herder, Joe Lucero, from Albuquerque. The sagebrush was thick, but we strung the sheep down the narrow dirt road and I got an accurate count. My count was 3,738, and I was super happy, but didn't let on. I acted most serious, and said, "Joe, we gave you 3,740 the first of May. Now I count only 3,738." Joe replied that he found one dead sheep, and I answered that he was still out one. He thought for a moment, then told me that he discovered a yearling wether in this herd of ewes, and butchered it for meat. I smiled and shook hands with Joe, told him he was doing a great job, and said we appreciated it.

Between the two herds, close to 7,500 yearlings, we were short only six, and that was wonderful. We were having fewer coyote problems than usual, and the men were good herders. Things were progressing, lambing and docking were completed, and shearing was well underway. Visitors came every day

Still from Encyclopedia Britannica film From Fleece to Fabric.
George Rasmussen, on his saddle horse, and his border collie heading out to gather his herd and take them to the shearing corral.

to see this great shearing process. Encyclopedia Britannica sent a team to film our shearing. They wanted to make a short educational film, *From Fleece to Fabric*, for use by universities and other educational institutions. I helped make the arrangements, and they started filming a herd of 1,200 ewes, plus their lambs, grazing on the hills five miles north. This was the herd looked after by George Rasmussen. They filmed him as he came out the door of his camp-wagon, mounted his white saddle horse, and proceeded to gather the sheep with the help of his dogs. It then showed a herd heading in to be shorn.

The film crew did a professional job throughout the entire process. They filmed the herd entering the sorting corral, where Will and I separated the lambs from their mothers before the ewes entered the shearing shed. Then the film concentrated on the long line of twenty-two shearers hard at work doing the actual shearing. They finished at the Deseret Live Stock Company, where the wool was sacked, and then transported to the railroad and shipped to a mill in the East. That evening I handed a dollar bill to Kathryn, and told her that was my pay for being in the movie. Encyclopedia Britannica paid each of us one dollar, and we signed a release so we could be shown in the film. We later saw the entire film, which included the manufacturing process and ended up showing the finished fabric. We were pleased. The film was successful, and several hundred copies were distributed throughout the United States.

One day, the president of Deseret Live Stock Company, Henry D. Moyle, came with several of his family members to view the process, and he brought his own movie camera for recording the action. Later, he and I visited about several things. He said he had heard some good reports about me and was glad to have me on the outfit. Finally, he said he had been told I was extravagant. I responded by saying that must mean I was spending too much money for something, or using more of something than was necessary. He said that that was what he had been told. I replied that the only place where I could have been extravagant was on the grub orders we supplied to the men out in the camps. When I came in June the previous year, our sheepherders and camptenders were well fed by most standards, but were still being told that because of shortages during the war, jam and pickles were just not available. Yet when they brought their herds in for shearing, and they ate here at the shearing corral, it was like Thanksgiving. There was plenty of jam and, of course, everything else. That situation didn't go over very well.

He just listened, so I told him how Will Sorensen and I went over the grub orders last July, when I was told to strike out requests for jam and pickles. When these items were not included in their grub orders, one of the men jumped on me about it and said that if the company didn't want to supply these things, then we should say so, but not tell them they weren't available. President Moyle asked, "Are you through?" and I replied, "Almost." I told him that Mr. Dansie, Will, and I had a three-way conference, and I asked how they wanted me to handle this deal about jam. They said they wanted the men to be well fed and satisfied, and we figured six jars of jam for six weeks was alright. I added that I was sorry if Mr. Dansie said I was extravagant. President Moyle answered that it was not Walt Dansie who said anything about this, and that everyone agreed that they wanted our men to feel good about working for the company, since we depended on them for our profit. He told me that he was glad he and I had a chance to get this situation straightened out.

We shook hands. All was well. Then he asked how often Kathryn and I and children went to church. I replied that we had not been to church since I came to work over a year ago, but that Kathryn and our kids went to church when they were in Manti last fall, during shipping time. He put his arm around my shoulder and told me that on Sundays he wanted us to be in church. He thought this ought to be a high priority for our families, and wanted me to handle Deseret's livestock business so our family would make it to church. I told him that we would do better.

That evening, Kathryn and I were discussing how to make it happen. I told her that Henry D. Moyle was a tower of strength and purpose. He and I had a great discussion, and we shook hands again before he left. Kathryn and I felt good about being with the Deseret Live Stock Company. Next morning, I told Will that President Moyle wanted me and my family in church on Sunday. Will replied that that was what they used to tell him and Ralph Moss when they were young. Ralph always said that they were sent out riding so far on Saturday, it took all day Sunday to ride home.

Bill Green, the north rider, did not come back to work for the Deseret Live Stock Company after he was released from the hospital, although his job was kept open just in case he wanted to return. Near the end of shearing, Mr. Dansie, Will, and I conferred about a new north rider. I said that we could get through the summer if I just assumed the north rider's responsibility along with my own. By the time we shipped lambs in September, there might be some person on the outfit we could move into that job. Shearing the 40,000 sheep was completed on July 2, 1947. After the old cull sheep were sent to Omaha, it was time to move the twenty-eight miles to summer headquarters.

WESTERN HOSPITALITY

IN JULY 1947, IT WAS a marvelous, exhilarating experience to be back on the gently rolling ridges of the 9,000-foot-high summer range, fragrant with wildflowers, conifer trees, quaking aspens, and abundant grass. Kathryn said this was beautiful, but the living conditions were not easy for a family with two small children and expecting another. She was right. I told her we could handle this situation for now, and things would get better for us as time went on. That summer, I did the work of the north rider in addition to being foreman over the whole sheep operation. I saddled up a trusty horse a couple of days each week and rode alone for most of the day, inspecting the ranges of the eleven herds of ewes and lambs on the north end. Whenever it was possible to go in the pickup, Kathryn and the kids rode with me. It was a real pleasure to visit the sheep camps, and let our family get better acquainted with the herders and camptenders.

Will Sorensen lived in his cabin at summer headquarters and ate breakfast and supper with our family. He spent part of each day supervising a Caterpiller, doing road work and water development. He also worked with the six-man fence crew, camped two miles away. Warren Cushman, a husky six-footer in his midthirties, with thick brown hair, a ready smile, and an easygoing disposition, cooked for the fence crew. Warren was an expert at baking bread and biscuits, and his breakfast hotcakes were a work of art. The boys on the crew worked hard, and I made sure Warren had a wide variety of vegetables and fruit, in addition to mutton, potatoes, ham, bacon, eggs, beans, breakfast cereals, and many other food items.

One day, when Kathryn and the kids rode with me in the pickup en route to Wahsatch, I stopped the truck near Chicken Spring, so Bill and Diane could observe two separate flocks of grouse crossing the dusty road. Chicken

103

Spring was well named, and the thick groundcover of shrubs and grass surrounding the pond helped nurture many broods each year. Near Wahsatch, as we drove alongside a small lake, the sudden appearance of the truck frightened a few ducks, which flew off the water. A hawk was perched nearby, waiting for this opportunity. It immediately took off and attacked a duck in midair, sinking its talons deep into the duck's back. We all saw the duck flutter to the ground about thirty feet from the truck. The hawk, in hot pursuit, landed and attacked again, using its beak to pound on the top of the duck's head.

I said, "Maybe I can save that duck," so I stopped the truck and walked toward the two birds. When I was about fifteen feet from them, the hawk flew a short distance away. The duck was groggy from the blows to its head, but it staggered and waddled helplessly toward me, and stopped between my feet. Blood was oozing through the feathers on the duck's back from where the hawk's talons had been imbedded. I picked up the duck and spread open the feathers. It was evident the duck could not survive those deep wounds, plus the injuries to its bleeding head. I didn't want the duck to suffer a prolonged death, which it would if it was gradually killed by the hawk. Also, if I took the hawk's prey, it would have to make another kill. I took out my knife and in a split second, mercifully, off came the duck's head. Its wings didn't even flutter; it was out of its misery. After washing the blood off my knife and hands, I got back into the truck. The hawk was already picking at the dead duck.

What we saw was a good example of what happens out in the world of wild animals. Some birds and animals survive by killing other birds and animals. Sometimes it takes several minutes, or maybe an hour, for one animal to kill another, and sometimes the stronger animal starts eating the weaker before it is dead. Living an isolated life far out in the mountains, we witnessed the reality of life and death as Nature's wild creatures struggled to survive. In this struggle, death comes to the weaker, which may be neither old nor sick. It may be a newborn nursing its mother, or it may be in the prime of life, just weaker than the attacker, or attackers. Death in the wild does not come easily.

Beavers were plentiful over much of the summer range. They were unmolested by trappers, so they thrived, multiplied, and built hundreds of dams along the streams. The cowboys and shepherds protected the beavers, because their dams prevented floods and erosion, and the beaver ponds made it easy for cattle, sheep, deer, and elk to drink. These ponds were alive with trout.

One day I was walking down a vague old wagon road, alongside a small stream about a mile below our cabin. The grass was thick and tall. Sheep would not graze here until late summer, and the cattle usually didn't venture this high on the summer range, because of an abundance of feed and water in the lower elevations. My purpose on this walk was to check the condition of the stream banks, which were covered with thick vegetation and showed no erosion. The water was clear, except where beavers were working. As I came alongside a beaver pond and was looking at the beaver lodge near the far side of the water, about forty yards distant, the ground seemed to explode at my feet. I had nearly stepped on a litter of young beavers sunning themselves. I didn't see them until they leapt from under my approaching foot and scrambled the twenty feet to the water. They were about the size of a large housecat, with a wide, paddlelike tail and light brown fur. I counted five of them as they swam on the surface of the water for thirty yards, paddling furiously with that characteristic beavertail movement, and then suddenly disappeared under the water to the entrance to their home. The whole episode was over in a few seconds.

I paused for a moment reflecting on the beauty of this situation. Here was a pristine setting, with groves of quaking aspen scattered over the ridges, particularly on the sunny southern slopes extending to the water's edge, and pines, firs, and spruces on the more shaded north slopes. The thick undercover of vegetation provided luxuriant grazing for sheep, deer, and elk, and it was a protective habitat for wildlife, particularly for the young.

Kathryn and young Bill, now nearly 5, and Diane, nearly 4, left the ranch and went to Manti in September 1947, so Bill could attend kindergarten while Kathryn awaited the birth of our next baby. Although Kathryn and the kids stayed with her folks most of the time, both sets of grandparents enjoyed having them visit. There was a hospital at Gunnison, fifteen miles south of Manti, and one at Mt. Pleasant, twenty miles north. However, there was no hospital in Manti. Kathryn was to have the baby at the home of Dr. Lucian Sears, our family doctor in Manti. His wife, Sylvia, was a nurse, and they wanted Kathryn to be at their home. They fixed a special room for Kathryn. This was a beautiful and considerate thing for them to do. Dr. Lucian's father, Dr. George L. Sears, had been the main doctor in Manti for many years.

After the fence crew disbanded at the end of August, Tom Judd cooked at summer headquarters after Kathryn and the kids left for Manti. Tom was single and had never been married. He was an excellent cook and made particularly good homemade bread. When the two herds that grazed part of the summer on the Cache National Forest were brought back inside the company fence, they were corralled so neighboring sheepmen could look through the herds for possible strays. One neighbor, Nick Chournos, after eating one of Tom's dinners, told me that we were lucky to have a man like Tom.

Joe Manzaneros had done a good job for the company; however, he had not returned to work after taking considerable time off. In his absence, Paul Silva from Ignacio, Colorado, was hired as the south rider. He was an experienced man, having served as foreman in charge of ten thousand sheep for the Ute Indian Agency in southwestern Colorado. Paul was a key man, trim and slim, who reminded me of Errol Flynn. As usual, the south-end lambs were cut out, trailed to Wahsatch, and shipped east in mid-September. When the north-end lambs were processed in September, I conferred with Will, and then asked Eldon Larsen, one of our herders, to be the north rider. Eldon was a single young man in his midtwenties, a longtime employee. He had worked here before serving in the U.S. Army in Europe during World War II. He had been in several bloody campaigns and was one of a very few survivors of his original outfit. He was slender, rode tall in the saddle, and knew the summer and winter ranges.

At the third and final lamb shipping, the last of September, about a thousand lambs were going to be held overnight in the Union Pacific stockyards at Wahsatch. Concerned about the safety of the lambs, I had two extra campwagons pulled alongside the corral. Eldon Larsen and Paul Silva slept in one wagon, and I slept in the other. During the night, I was awakened by the trains being switched and rail cars moving along the siding for shipping the next morning. I got out of bed, pulled on Levi's and boots, and slid open the camp door to take a look. Suddenly the huge steam locomotive on the tracks next to the corral let off a big white cloud of pressurized steam in the direction of the lambs. The escaping steam's loud whishing frightened the lambs, and they stampeded to the end of the corral near the campwagons. They hit the heavy wooden plank fence and piled up against it, three or four deep. By now the steam from the engine had stopped.

I opened the door on the other camp and yelled, "Get up and come help me." I said that we had lambs piled up in the corner, and they would smother. We climbed into the corral and started unpiling lambs. It was like a football pileup, except there were about thirty lambs that could smother. We worked furiously. It took only about a minute, certainly not much longer, and then all the lambs were standing. This could have been death to many lambs. I went over to the train and talked to the embarrassed engineer, an amiable, stout man. He removed his striped cap, wiped his brow with a red handkerchief, apologized, and told me that it would not happen again. He didn't realize lambs were so spooky.

We men at the Deseret Live Stock Company had a good working relationship with the Union Pacific railroad men, and it was essential to keep everything on a pleasant, cooperative basis. I told the engineer that the Union Pacific crew always had our cars on hand before we were ready to load. If we specified forty-two double-deck cars, they brought forty-four, just in case the extras were needed. Over the five days when we shipped from Wahsatch to the

Sheep going through the dipping vat at Salaratus Creek, four miles south of Home Ranch headquarters, October 1947. Will Sorensen is in the foreground.

winter range, we always had the whole train to ourselves each day. The same was true when we shipped from the winter range back to Wahsatch. We were truly accommodated at all times. After hearing that, the engineer felt better.

In late September 1947, after the lambs were shipped, Will Sorensen supervised the building of a dipping vat, log corrals, and a cookhouse. They were alongside Salaratus Creek, at the spring near the mouth of Black Dan Canyon, about four miles south of Home Ranch headquarters. This big spring provided plenty of water. Walter Dansie, Will Sorensen, and I had studied the situation and figured dipping the sheep would get rid of all their external parasites, such as ticks, or "keds," as they were sometimes called. Also, a corral at that location would facilitate making up the winter herds in October, at dipping time. This would spare the range around the shearing corral by keeping the sheep off the rolling, grass-covered hills around Wahsatch until just before shipping in November.

Dipping the sheep in cold water in October would be chilling. Will Sorensen said that in order to warm the water in the dipping vat, we could install a 3,000-gallon tank between the spring and the vat. Then water piped from the spring into the tank could be heated and piped to the vat. We set the water tank on steel railroad rails and removed sufficient earth so a fire could be built under the water tank. Two men hauled truckloads of slabs from the sawmill in Peck Canyon, seven miles away. We started the fire at six o'clock each morning and burned those slabs under the tank to heat the water. It worked.

The three-room cookhouse had a kitchen/dining room that would accommodate sixteen people at mealtime, a bedroom for the cook, and a storage room with shelves for food. The walls were made of sawn logs, and the floor of regular tongue-and-groove lumber. A large black kitchen stove burned both wood and coal. Emma Zabriskie came to cook again that fall, and of course everyone enjoyed her meals.

While building the corral, Will Sorensen accidentally got a heavy fence log rolled onto one foot and ankle. He couldn't walk on it, and it swelled up and turned black and blue. He wouldn't go to the doctor. When the foot showed no improvement after a week, I told Will that I wanted to take him to the doctor and have the foot and ankle x-rayed. He refused, saying that it was a contest to see who was the toughest, him or the foot.

The work progressed exceedingly well. Putting sheep through a dipping vat is one of the least pleasant jobs in the business. Each animal must be completely immersed in the water, which contains the proper amount of insecticide. The insecticide kills all external parasites, and the sheep emerge "tick free." After our sheep climbed out of the vat, they were held in a dripping pen with a solid wooden floor, sloped to allow the water to run back into the vat. A fresh paint brand was applied to all the sheep, while they were still wet, by Eldon Larsen. He said that the wet brands were not quite as sharp as when applied to dry wool, but if we could read them, then so could the herders. We also accomplished our other main objective, keeping the sheep off the much-used range around the shearing corral and near Wahsatch.

When dipping was completed, we moved the eighteen miles over to shearing headquarters. In late October, final jobs were completed in preparation for shipping to the winter range. The nights were freezing, and the mornings cold enough to require a heavy coat. The sun usually warmed up the countryside by noon. When a stiff wind blew across Wahsatch, there was no doubt summer was long gone. Emma rang the first bell every morning at six o'clock. Everyone was ready for breakfast at six thirty, and those were sumptuous meals. Usually she served hotcakes or hot biscuits, hash brown potatoes, ham or bacon, eggs, hot cereal, fresh fruit, and a choice of beverages.

At Wahsatch, we again shipped fourteen herds to the winter range. We started November 1, and sent three herds each day for four days, loading the last two herds on November 5. Late that afternoon, Emma Zabriskie rode to Salt Lake City with Mr. Dansie, and Will Sorensen and I drove to Skull Valley.

11

A New Addition

I was awakened at 5:00 a.m. by loud knocking at the door. When I opened it, Glen Hess told me that Kathryn had just phoned, and that we had a new baby boy, born November 5, 1947. She said everything was O.K., and that the baby weighed nine pounds. I thanked Glen, and told him I would call Kathryn later that morning. Now wide awake, I was thankful and happy. We had decided on the name Dale Conrad if it was a boy. When the crew gathered over at the cookhouse at six thirty for breakfast, there were some of the same old cowboys and ranch hands at the table, and a couple of new ones were introduced. Each said congratulations on the baby. It was a great day. When I called Kathryn, we decided that I would go to Manti in a few days and pick them up. It was good to get the family together again, and all were pleased to have a new baby boy. Everybody was strong and healthy, and we were a happy family. We enjoyed visiting parents, brothers, and sisters in Manti.

Back in Skull Valley, all the herds were located on their designated ranges. Feed was plentiful, due to early fall rains. The sheep and horses filled up and stayed fat. Herders and camptenders were happy. In the evenings, I enjoyed holding Dale and rocking in the chair while Kathryn read to Bill and Diane. I told Kathryn that I was pleased with all the good books she bought for the kids. She had always done a lot of reading.

We distributed fresh beef and pork to all of our camps in November and during the first week of December. The next big job was to put rams into the ewe herds. On December 14, the south-end rams were driven to the corral near White Rock. I wanted to see first-hand how driving the rams over Cedar Mountain was working. So on the fourteenth I rode my big black horse, Butch, and drove the north-end rams across Skull Valley, put them in the corral at Eight-Mile for the night, and stayed at the closest camp. The next morning, we separated rams for the two herds in that area, and I drove the remaining 240 rams, enough for four herds of ewes, up the canyon west of Eight-Mile, and headed for the top of Cedar Mountain. There was no road, just a dim horse trail. It was a bad operation. The snow was much deeper than expected, and the rams moved at a snail's pace. Noon came, and midafternoon, and four o'clock. They were still climbing, but progress was slow, and the rams were tired. At dark, I had to leave the rams still on the east side, not yet to the summit. They were "give out."

I knew Will Sorensen, Austin Christofferson, and Eldon Larsen were waiting for me and the rams to arrive at the north corral, on the west side of the mountain. I rode on over the top, and there, far below, I could see a fire burning where they waited. They were surprised when I arrived at the corral way after dark, without the rams. I told them the snow was deep, and the rams couldn't travel farther that day; they were on the east side, about a quarter mile below the top. Will said that Austin had better stay with Eldon for the night, then take my horse and bring those rams down here the next morning. Those two, and the camptenders, could separate the rams, and the herders would all get the rams the next day, only a little later than expected. That was the last time we drove the rams that distance. From then on, they were trucked.

Mr. Dansie brought out the Christmas checks, and we got them distributed to each employee by December 20. As usual, most of those checks were immediately enclosed in letters to families. Will Sorensen wanted to spend some of the holidays with a niece and her family in Jackson Hole, Wyoming; he was gone for two weeks. Kathryn and I enjoyed Will, so we missed him. Our family had a great Christmas.

The National Woolgrowers' Convention was going to be held in Salt Lake City, at the Hotel Utah, from January 25 through 29, 1948. Mr. Dansie told me that I needed to attend, and was to take Kathryn and our three children. He made reservations at the hotel, so we could take care of the family and also participate in the meetings. I again served on the Lamb Marketing Committee, having been assigned to that committee by the National Woolgrowers. This was a well-attended convention, the meetings and program interesting and educational.

At that convention, Farrington R. Carpenter of Hayden, Colorado, was a featured speaker. He told about his experiences organizing the grazing districts on land owned by the U.S. government. This was a result of the Taylor Grazing Act, legislation proposed by the ranchers, which brought about federal control of grazing on those public lands. "Ferry" Carpenter was the first director of what is now the BLM (Bureau of Land Management), and

Ed Solberg, Will Sorensen, and Dean Frischknecht, rear, left to right and Diane and "Bill" Dean Frischknecht in front, building a garage with discarded railroad ties, to house five vehicles, spring 1948, Skull Valley.

he had the responsibility of adjudicating numbers of livestock permitted to graze on each grazing allotment, and issuing the permits to stockgrowers. Mr. Carpenter spoke for an hour, the rapt audience stood up and stretched while he got a drink of water, and then he spoke for another hour. He said that the three preference categories in establishing who was permitted to use the range were: category 3, those who were near; category 2, those who were nearer; and category 1, those who were nearest. Nearest usually prevailed. He was a knowledgeable, dynamic, and entertaining speaker.

Afterward, I took time to visit with Mr. Carpenter but mainly listened. He told me that he grew up in Chicago, where his dad was a businessman. He graduated from law school at Princeton University, and as a student was closely associated with Woodrow Wilson, who was then president of that institution, prior to becoming president of the United States. Upon graduation, he homesteaded near Hayden, Colorado, creating the base for a cattle ranch, and practiced law. This chat was the beginning of a long working relationship with Mr. Carpenter.

That winter, Will and I decided to build a low-cost garage to house the trucks and an auto or two. We gathered discarded wooden railroad ties along the Union Pacific tracks and used these to build a garage, big enough to house five vehicles, at ranch headquarters. We installed a stove made from a 50-gallon barrel, and windows were strategically located.

Paul Silva on a Palomino horse and Will Sorensen holding his favorite horse, Curly. Will is having a conference with Paul as they prepare to ride out from the shearing corral.

A few days later, at Luis Zozoya's camp on the west side of Cedar Mountain—at the extreme south end, near Wig Mountain—I noticed his saddle mare appeared to be pregnant. She was getting a larger belly than usual, and her udder was starting to develop. Luis said that last April, she was hobbled out most of the time, and strayed no more than a half mile from camp. The wild mustang stallion he saw hanging around must have bred her. His camptender brought the pregnant mare to the ranch, and took out a fresh horse for Luis. The mare gave birth to a nice-looking foal, and they were left at the ranch.

Paul Silva and I needed a corral on the west side of Cedar Mountain, on the south end. We wanted a smooth-working chute. I purchased the wire and lumber and hauled it to Paul's camp at White Rock. However, the steep, rough road to the desired location meant we couldn't haul the long lumber in a truck. A resourceful camptender, John Arellano, came with his team and wagon and hauled the lumber. He put his log chain around the front of his wagon-box, then used the wagon-jack, braced against the chain, to keep the lumber riding level. We built a strong corral on private land owned by the company. It held 2,800 sheep, and the cutting chute worked excellently, as expected. From November 1947 through April 1948, we had the best winter and the lightest sheep losses ever experienced by the Deseret Live Stock Company up until that time. The winter feed had been plentiful, and coyote losses were light. Shipping off the desert went smoothly during the last week of April. Wahsatch was a magnificent sight, reservoirs full, feed plentiful, the sun shining. Emma Zabriskie came back to Wahsatch to cook during lambing and docking. She added enjoyment to each meal.

Bill Gustin, holding a big lamb we missed at docking; Paul Silva, standing in
the mouthing chute removing testicles and tail; and Fred Martinez ready to put
a lamb brand on a big lamb, spring 1949.

That spring, May 1948, was the best lambing in the history of the Deseret
Live Stock Company. Docking counts showed 31,237 lambs docked, a new
high. In two locations, 250 ewes had drifted into another herd. I didn't like
it, but had to leave them undisturbed. This would cost us some lambs. Will
told me this was the first time we'd had twenty-two herds of ewes and lambs.
Usually it was twenty herds with lambs, or twenty-one herds on a good year.

Shearing started June 15, and again Bert Robbins's crew from Santaquin,
Utah, did the job. Mrs. Robbins had previously arranged for Emma to stay
and help with the cooking. She also brought an attractive young lady, Beth, to
help in the kitchen, set the tables, and keep the bowls and platters full at meal
time. These three women did a remarkable job, feeding fifty men three huge
meals each day.

June 20 was the day cowboys from the Home Ranch distributed bulls
across the country, from the Home Ranch to the shearing corral. Ralph's sons,
John and Bob Moss, brought thirty Hereford bulls to stay overnight at our
cattle corral. Their horses were stabled in our horse barn. Cows and calves
were grazing on south for several miles, and these bulls would be distributed
on that range. Mr. Dansie and Ralph Moss deserved much credit for buying
high-quality bulls. Our outfit raised "reputation cattle."

Back at the house, Kathryn reminded me that this was where Clarence and Marie Rasmussen met, when he was the south rider and she came to help with the cooking during shearing. They were a happy couple. Now it was possible that our north rider, Eldon Larsen, and Beth might strike up a romance. She was a fine young lady, and he was a good young man, so that would be great for both of them.

During the first week in July, we completed shearing 40,055 sheep. Mr. Dansie wrote a check to each shearer and all the temporary employees. It had been a successful lambing, docking, and shearing. Six husky high school boys, a man cook, and Bill Watts, as foreman, would be the summer fence crew. Will, Kathryn, and I helped close up the spring headquarters, and moved the twenty-eight miles back to the top of the summer range.

1948 Shearing counts to mountain.

Brand	Herder	Ewes	Lambs
R +	Ortega	1011	1035
R H	Arvel	1059	1101
R 3	al	1572	1746
R 1	Manuel	1233	1333
R 5	Dupre	1202	1432
R O	Willie	1239	1404
R θ	Frank	1117	1249
R 2	Cordova	1160	1260
R 10	Trimble	1376	1635
R V	Solomon	1220	1290
R N	Waldo	1317	1404
R T	Arellano	1484	1460
B 3	Gabino	1338	1455
B O	Pedroncello	1013	1097
B 5	Atencio	1270	1460
B 1	Tim	1148	1177
B +	Luis	1240	1243
B 2	Chavez	1444	1457
B 3	Trujillo	1314	1471
B V	Archuleta	1446	1549
B 10	Jaramillo	1431	1678
B H	Abeyta	1328	1452
		27,962	30,388

Dries-yearlings 10,781

38,743 ewes

after spring culling.

12
Timely Moves

Will told Kathryn and me that he was going to town for a couple of weeks. He had eaten most of his meals, particularly breakfast and supper, with us for the past two years, and we missed him. Also, Austin Christofferson recently decided that he had enjoyed the far-out country long enough, and returned to his hometown, Spring City, to live. This left Kathryn cooking only for our family, except when company came. When Will returned, he was towing a new mobile home, and had an attractive gray-haired lady at his side. He introduced her as his wife, Vera. Kathryn and I were surprised, and told them we were happy for both of them. Vera said that Will had told her so much about our family, she was glad to finally get to meet all of us. She told us that all her life she had heard about the Deseret Live Stock Company. Now she would learn what it was to live here in this beautiful area.

We learned Will and Vera had known each other in Spring City for most of their lives, and were long-time friends. She had been a widow for years, so this was going to make life more pleasant for them both. Vera had an easy, outgoing personality. Not only was she good for Will, but Kathryn now had a woman to visit with. They located their mobile home a couple of hundred yards south of the cabins, in a grass-covered area adjoining a grove of aspens, fifty yards off the road. Vera was an excellent cook, and she and Will ate in their home. We had them over for dinner a few times, and Vera was well-tuned into caring for children. She was a great addition to the Deseret Live Stock Company.

Will checked closely on the fence crew as they repaired the existing fence and then built a few miles of new log-and-block fence. The company objective, to have a strong fence around the outside perimeter of the whole 250,000 acres, was progressing. My summer was going to be easier than the one last year, now that we had a north rider. Eldon Larsen camped alongside Tom Judd, half a mile south of headquarters. Eldon and I saddled up on several occasions and rode together to areas where a pickup could not go.

A short time earlier, Ralph Moss gave me two more fancy saddle horses, Snip and Redwing—1,200-pound half Thoroughbreds—with the instructions that I had better turn them over to one of our riders. I wanted them kept as a pair, so I turned them over to Eldon. After a couple of days, Eldon commented he was glad to have them, but Snip made him feel like he was settin' on a keg of dynamite. He couldn't fool around on Snip. On one of our rides, Eldon confided to me that his romance with Beth was the real thing. It was going to work out so they could get married in the fall, and when they got married, he was going to live and work in town. He was alerting me ahead of time that he would be leaving the Deseret Live Stock Company later in the fall.

Mr. Dansie drove to the summer range, and told us that he would be leaving July 27 for a two-week business trip to look over that land in Florida the company had been investigating. The president of Deseret Live Stock Company, Henry D. Moyle, had previously been there on other business. He had been told land was available south and east of Orlando, at a reasonable price. His informant, an experienced cattleman from the West, said Florida offered year-round grazing, and should be investigated. Mr. Dansie returned with a favorable report, and told us the contact man in Florida has been instructed to take options on this land. During the next few weeks, he obtained options on a block of 360,000 adjoining acres. The price was quoted at $9.00 per acre. Clearing off palmetto, building roads, fencing, fertilizing, and seeding would cost another $40 to $50 per acre. This Florida situation was looked upon by some stockholders in the Deseret Live Stock Company as a real opportunity; others were skeptical. Mr. Dansie and President Moyle studied the situation in detail.

Our son Bill was going to be six years old on September 27, 1948. Kathryn and I had talked this over with Mr. Dansie and Will some months ago, and everyone agreed that we should rent a house in Grantsville, so Bill could start first grade right after Labor Day. In late August, I moved Kathryn, Bill, Diane, and Dale from our high mountain home into a modest house rented from Denny Hale. It was two blocks north of Main, on a street the locals called "Cow Alley," a name kept on from earlier days, when many cattle were driven past. It was a convenient location. Another reason for wanting to move into town was that we were expecting another baby about Christmastime. Having the family in Grantsville, thirty miles west of Salt Lake City, was an excellent location. It was only thirty miles from Grantsville to the ranch in Skull Valley, and ten miles to the hospital in Tooele.

Back on the high range, lambs were to be marketed, as usual, in three large droves in September. All market lambs were trailed to the shearing corral, along with the old blue-dot ewes. That year the lambs were trucked to the stockyards at Ogden, Utah, where Mr. Dansie received them. There the fat lambs were separated from the feeders. The fats were immediately processed at a nearby packing plant. When the first lambs arrived in Ogden, Mr. Dansie sent a note to me via one of the eleven truckers returning for a second load. It read "Ogden—Sept. 13, 1948. 1,918 received so far. In the main they are coming in in pretty good shape. Let me know your program for the rest of the week. W.D." I consulted with the manager of the trucking crew and figured out how many loads it would take to haul the south-end lambs, as well as the number of days required. My written reply showed this information and concluded, "North-end lambs will not be ready to send to Ogden until after the 20th; maybe 21st or 22nd. I'll call you at the office when you have finished with these south-end lambs at Ogden."

When Mr. Dansie and I got together, he told me he had purchased 150 Romney rams from Coffin Brothers Sheep Company in Yakima, Washington. This was not 100 percent good news. I figured Columbia rams were coarse enough to cross with Rambouillet ewes in this brushy country. Will Sorensen had outgunned me on this deal. Mr. Dansie continued that he had purchased 90 Delaine Merino ram lambs from Ohio. He was not happy with our fine-wool Rambouillet rams. He thought their fleeces lacked density. He wanted to try a set of Delaine Merino rams on our herd of quarter-blood ewes. These Merinos were smaller in size than anything we had ever used, but Mr. Dansie said the ram lambs would be grown out under our conditions, so they should increase the density of fleeces in their offspring.

Now I had just heard two pieces of bad news. More Romneys, and now Merinos, and I had not been consulted. To me, this was extremely coarse wool that was going to be crossed with extremely fine wool. Genetically, this was a severe cross, which would take several generations to standardize. Some fleeces would have coarse fibers interspersed with fine fibers throughout the entire fleece. It had taken the U.S.D.A. at Dubois, Idaho, many years to create the Columbia breed by crossing the coarse-wool Lincoln breed with the fine-wool Rambouillet. Mainly, I was disappointed at not being in on the decision.

I was at work on the summer range when the Merino ram lambs were shipped by rail from Ohio directly to Timpie, then driven up Skull Valley to the ranch. They arrived before the rams were shipped from Wahsatch, and were put into a large feeding lot, but not into our main ram feeding lot. I made an overnight trip to Grantsville for a combined celebration of Bill's birthday September 27 and Kathryn's on the twenty-ninth. It was a wonderfully happy reunion. Kathryn was feeling good, and Bill enjoyed school. She said it was good to have inside plumbing and electricity, and all the people she met were going out of their way to make them feel at home. I made a quick run to Skull Valley to see the Merino ram lambs and stopped back at home before returning to the mountain. When Kathryn asked about the Merinos, I told her that they were too small. However, I was going to grow them out and give them an honest try.

While on the summer range, I offered Ace John, who was working for a neighbor, the north rider's job, succeeding Eldon Larsen. When it became convenient for the neighbor, Ace came to work later that fall. Will and I agreed that there was no need to dip the sheep in the fall of 1948, since the previous year's dipping still had them "tick free." It would be best to use the new corral at Salaratus to grade the herds and get fresh paint brands put on for winter. So we lived and worked there at the dipping corral during the middle of October, then moved to the shearing corral for the job of shearing eye wool.

Summer 1948 had been dry in much of the West. The feed on the winter range in western Utah was reported to be far short of normal. To get a first-hand look at Skull Valley and the range west of Cedar Mountain, Will Sorensen made a special trip in October. He was struck by the lack of forage. He knew I was in for big trouble, so he agreed to stay one more time. He told me we would have feed for sixty days, or once over the grazing country. He proposed to Mr. Dansie and the board of directors that we sell ten thousand sheep, rather than try to winter them. This proposal was given very serious consideration. Ultimately, it was decided to cull enough to reduce their numbers to the equivalent of a winter herd. We cut out 1,600 of the least-desirable ewes and 1,300 of the smaller ewe lambs, and shipped them to Omaha, a reduction of 2,900. We planned to leave one herd near the Home Ranch, and held all the herds back in the brushy country, away from Wahsatch, killing as much time as possible before shipping them to the desert.

In early November 1948, at the time of the general election, Governor Tom Dewey, the Republican candidate for president, was heavily favored to beat President Harry Truman, who was seeking reelection. President Truman had been elevated to the presidency from the vice-presidency, following the death of Franklin D. Roosevelt in April 1945. Many people thought Dewey would be an easy winner, and he appeared quite confident. The night of the election, some newspapers and magazines came out with the headline "Dewey Defeats Truman" before the votes were counted. After the votes were tallied, President Truman had decisively defeated Governor Dewey. Next morning at breakfast, Will didn't even smile. He just glared at me, since I had predicted a couple of days earlier that Dewey would win by three million votes. I said that all I was going on was what we heard on the radio from the newscasters.

A short time previously, I was in Grantsville with Kathryn and the kids. It was the night before Truman, from the back of his railroad car, was going to speak in Salt Lake City. I got my young son Bill up early, and we drove to the Union Pacific depot to hear him. Truman gave a sensible talk to a huge crowd. We were up close. I wanted my son to see a real live president of the United States, especially because Truman was favorable to people in agriculture.

I told Will I understood that when the president of the Deseret Live Stock Company, Henry D. Moyle, was seeking the Democratic nomination for governor of Utah a few years previously, Will hauled herders and camptenders to get them registered and then to vote. Will replied that he had hauled the men because Henry would have been a good man for the job. If he would run again, Will would work to get him elected. Will said he grew up being a Republican, because he was in favor of a tariff on wool to protect our sheep industry. Henry was a Democrat, but he knew we needed a wool tariff. I said that President Moyle's dad, James H. Moyle, who used to be president of the Deseret Live Stock Company, was under-secretary of the U.S. Treasury, and was appointed by Woodrow Wilson. They were a family of able people.

However, we had work to do. Will proposed that he would get the sheep shipped, if I would go west to receive them. When I got to Skull Valley, the Merino ram lambs, which had been kept separate for a month, had been put in with the ram herd. Some of those Merinos were limping. I examined one, and it had foot rot. I then telephoned Mr. Dansie. Dr. Osguthorpe, a veterinarian, came out, and I was glad to see him. We had taken classes together when students at Utah State. When he examined the first infected foot, he said, "You're right—that's foot rot, and you know it is very contagious." We would have to examine and treat the feet on every ram lamb. Dr. Osguthorpe said that the Merinos brought this disease with them. Soon there were several rams in the main herd showing signs of foot rot. I called Dr. Osguthorpe, and we outlined a foot-rot-eradication program. Two young men helped me build two new feeding lots for the rams. We dug a trench for a water line, and piped in drinking water. We caught every ram, sat each one on its rear end, and then Dr. Osguthorpe examined and treated each foot. We handled nearly nine hundred rams. About seven hundred of them, free of the disease, were separated and put into one of the new lots. The remaining two hundred were treated and put into the other new lot. Eventually we got the disease under control, but it took twice-a-week treatments for three weeks.

When Will and I drove out over the winter range in my new four-wheel-drive Jeep pickup, he said that we would have to hope for a mild winter. We'd have to "winter on the weather." The feed would last maybe two months. We trucked an emergency supply of two hundred bags of grain pellets and two hundred bales of alfalfa hay to the stackyard and granary on the west side of Cedar Mountain, where the north rider camped. Things were going well, and we hoped the winter would not be too tough.

It was a good thing Kathryn and I moved our family to Grantsville that fall, because in mid-November she was being threatened with a possible miscarriage. I was living in a house at the ranch and getting home overnight about every three days. One afternoon out at the ranch my friend Jess Charles drove up. He told me that an ambulance had transported Kathryn to the hospital in Tooele. His next-door neighbor, Fanny Anderson, was tending our three kids until I could get home. Fanny was a registered nurse and a wonderful neighbor. She had been helping Kathryn a great deal. I thanked Jess, then talked to Will. He said for me to go take care of the family, and he would handle the work. I said I would let him know how Kathryn was.

After a thirty-minute drive, I pulled up to our home in Grantsville. When I walked into the house, Fanny was sitting on the couch holding one-year-old Dale. Fanny said that she would tend the kids, and that I should go to Tooele and check on Kathryn. Kathryn had called her mother in Manti and asked her to come to Grantsville and help with the kids. Her mother left on the bus at once. I told Fanny that we were happy she was our close neighbor. We appreciated all the help she gave to Kathryn and the kids.

At the hospital, I learned Kathryn had arrived in time to prevent a miscarriage. Between hugs, she told me that the doctor said she would be flat on her back in the hospital for two weeks. I met Kathryn's mother at the bus station in Grantsville. I told her we were truly grateful for her coming to help us, and that the doctor said Kathryn would be all right. Now, although I had a bed in a house at the ranch, I went home nearly every night to check on the family and visit Kathryn in the hospital. She recovered rapidly and was allowed to return home by ambulance after a two-week hospital stay. At home, she still stayed flat on her back. It was a good thing her mother was there.

The cowboys and sheepmen formed a combined workforce for hog butchering and pork processing in early December. Before the rams were put into the ewe herds on December 15 and 16, Dr. Osguthorpe came out and checked each ram as it was set on its rear end. Forty rams were still questionable, so we trucked them to a packing plant to be slaughtered. It was not worth taking a chance and spreading foot rot into the herds of ewes. That foot rot was one nasty, expensive experience. We trucked out rams to ten herds on the fifteenth, and had the remainder ready to go the next day. Will said that we would know a lot more about the winter when we brought the bucks home at the end of January.

I was bedded down at the ranch that night when Bill and Gene Miller, close friends from near Grantsville, drove to the house where I was sleeping and pressed hard on the automobile horn. I went to the door. Bill told me Kathryn had gone in an ambulance to the hospital in Tooele for the birth of our baby, and she wanted me to come. I thanked them. I got into high gear, drove home to Grantsville, cleaned up, changed clothes, and pulled up to the hospital in Tooele in plenty of time. Kathryn and I had a good visit. All went well. Doris was born December 16, 1948. She weighed nearly seven pounds, and was fully developed and strong. We were a very happy family, with two boys and two girls. I made it back home, and to the hospital, the next two nights. Things were going well for Kathryn and the new baby, and for Grandma Sorensen and the children.

On December 19, Mr. Dansie came to Skull Valley with a Christmas check for each employee. Again, it was for one-twelfth of their earnings during the year, a month's wages for full-time employees. This was a wonderful thing. Many of our men had letters ready to go home, and the check was included.

Kathryn and the baby would be released and home for Christmas. I thought she would enjoy a new rocking chair, so bought the best one available in Tooele. There was over a foot of snow, and when Kathryn and the baby were ready to come home, I drove the four-wheel Jeep pickup right up to the front door of the hospital. It was good to have the family together again. Christmas in Grantsville was a happy time. Kathryn's mother planned to stay and help until after Christmas, and Kathryn's dad came from Manti, so the two of them would go home together.

The day after Christmas, as I drove up Skull Valley, I could see a high column of white smoke ascending from the ranch. When I got there, all that was left of the big old ranchhouse were a few smoldering embers. It had just burned to the ground. It was the cookhouse, it had a large dining room where the crew ate, and was home to Glen and Edna Hess. The burned skeletons of the metal beds and kitchen equipment were all that remained. Glen did a good job of improvising. He told me he had to take the house where some of us men were sleeping, and use it as the house for cooking and eating. This house, although not as spacious as the one that burned, worked just fine. We men moved out into an older home, unoccupied for years. It had four rooms, a roof above, and a stove. All the burned debris was cleared away the next day. Mr. Dansie drove up and, after a conference with Glen, said he would have an architect come out and draw up plans for a new, modern house. He was sure we could get going on it as soon as the weather moderated in early spring.

THE BIG SNOW

THE WINTER OF 1948–49 WAS the winter of the "big snow" throughout much of the West. It was my third winter as sheep foreman for the Deseret Live Stock Company. The feed on much of the winter range was extremely short, and to make matters worse, too much snow was piling up. It was a good thing Will Sorensen had married Vera and decided to stay on and give me counsel and help as needed. Will had parked their mobile home at the ranch in Skull Valley. He and Vera were now living in the house where Kathryn and I had lived before moving into Grantsville.

On January 13, 1949, I invited Walter Dansie and Will Sorensen to ride with me in the four-wheel-drive Jeep pickup truck and look over the situation on the west side of Cedar Mountain. We saw two herds of sheep near the north end, and although snow was over a foot deep, the sheep looked good. When we got out to shake hands and visit, the herders and camptenders told us they needed supplemental feed. We explained that we were there to look things over and then decide.

I then drove in a westerly direction, down to the flat on the east side of the old bombing range. We wanted to see a herd of sheep that was west of Rattlesnake Point, eleven miles on south. As we drove along the east side of the bombing range, on flat country, the snow was twenty inches deep, and the front bumper was pushing snow as we traveled. There was no road, or even tracks, to follow, but I determined the approximate location of the camp. I took a firm hold on the steering wheel, and braced my left elbow on the door, in order to keep heading straight for the camp. A bank of fog drifted in; visibility was only twenty feet. The course had been set, we were traveling in a straight line, there were no rocks or washed-out gullies, and the country was flat. Progress was visible on the speedometer. We soon covered ten miles. Still in a dense fog, proceeding slowly, I knew we would soon be to the sandy hills. The fog began to thin out, with visibility about sixty yards ahead. Suddenly we drove right into the tracks of the wagon and followed them three hundred yards to the camp.

The camptender, Cornelius, was a large man. He was wearing blue bib overalls, and when he stepped out of the camp, all of us shook hands and extended greetings. I said we wanted to see how the sheep were doing, and to say hello to Candido. Candido had the sheep out west just a few hundred yards, in the sand dunes, so we could walk to see them. Within three hundred yards, we saw 1,000 of the 2,600 sheep in the herd. Candido came on his horse. His two dogs started barking at the strangers, and he had to quiet them. I greeted him and asked how it was going. He shook his head and said, "Bad. It's bad." He was a small, trim man and seemed almost to float as he stepped down from his horse and shook hands all around. He said a lot of sheep would die if they didn't get some feed. Mr. Dansie told both of them that we came to talk, see the sheep, and then decide what was best. The sheep were not getting enough to eat, but they were still strong. We discussed the situation while walking back toward camp. Then Cornelius said, in his gentlemanly manner, that he had just baked bread and had a big pot of pinto beans and ham ready, and that we had better eat something before heading back. We did, and it hit the spot. Just before getting into the truck, Cornelius pulled me aside and said, very firmly but in a soft voice, if we weren't going to feed those sheep, we should get a man to take his place.

As we three decision-makers headed back north, through twenty-five miles of deep snow, to Clive on U.S. Highway 40, we discussed the situation from every angle. Walter Dansie said we hadn't ever had to feed all 40,000 sheep, and we really weren't equipped to feed all those herds. Will reasoned that the days were getting longer, and by February 10, the sunshine in the afternoon would cause the snow to melt. My senior partners indicated that I would have to tough it out through the next two or three weeks.

Our decision that day was not to get supplemental feed to the main herds of sheep. We were already feeding hay to the herd of thin sheep, about 1,500 ewes, just east of the ranch in Skull Valley, and were feeding grain pellets to the one herd of aged ewes north of the ranch. I told them that we had ten tons of grain pellets, that is, two hundred bags of emergency feed, in the granary at the north rider's camp, on the west side of Cedar Mountain, and two hundred bales of alfalfa hay in that stackyard. I thought we had better double that supply. Walter Dansie said he would order more pellets to be delivered to the ranch and would order a railroad car of corn to be put on the siding at Timpie. He would call us when the corn was to arrive. It had been a day of difficult decisions.

We began hauling emergency feed not only to the west side of Cedar Mountain, but to our storage place on the south end in Skull Valley. Mr. Dansie supplied us with a new one-and-a-half-ton farm truck with a high stock rack. The heavy snow had drifted. Some days we shoveled through drifts three feet deep and thirty yards long. We had the big truck and my Jeep pickup loaded with feed. It worked best if the big truck followed the Jeep. When we came to a bad place, I walked to determine the situation. Sometimes the Jeep would get high-centered on the frozen snow, and all four wheels would spin in the air. The winch and cable from the front end of the big truck pulled the Jeep back out of the drift. Then we shoveled.

Days were bitter cold. Wind varied from a slight breeze to sharp and stinging. Already tanned faces turned a deeper brown from the sun and wind. Sunglasses or goggles were necessary to cut down on the constant glare of the sun on days when it wasn't snowing. Each man used light cotton gloves under larger leather gloves or mittens. In my pickup truck, I carried a large, navy blue wool overcoat given to me years earlier by my grandfather, Conrad Frischknecht, Sr. The coat was heavy and extra long, reaching almost to my ankles. Grandpa used it during the winter when driving his horsedrawn buggy in Manti. As it turned out, I had several occasions to use that coat.

I told Will we needed to prepare an emergency box to carry in each truck. We fixed a box of food items, a first-aid kit, matches, and a small bundle of dry kindling. Normally each truck carried an axe, shovel, big jack, tire chains, flashlight, pliers, and other tools. Each truck now also carried an empty coffee can filled with sand that was dampened with a cup of gasoline mixed into the sand, with the lid put back tightly on the damp sand. This can of sand, stored in a box behind the seat, was a safe and sure source of starting a fire in case of emergency. I explained that it was not necessary to use the whole can of sand; a cupful would do the job of helping to get a fire started.

During late January and early February, we struggled to get hay and corn or pellets to strategic locations in Skull Valley and on the west side of Cedar Mountain. Then it began to snow again. At the ranch in Skull Valley, we awoke one morning to another sixteen inches of new snow on top of fourteen inches of old snow. I called Mr. Dansie in Salt Lake City to advise him of the situation. He said he would check to see if the county or some emergency agency was going to open the road up Skull Valley. Soon he called back to say a "Cat" from Tooele would open up the road that day. The Caterpillar made it to the ranch at four thirty in the afternoon. An alternate Cat driver followed, in a pickup truck carrying diesel fuel. When they reached the ranch, they plowed a road to the nearest stackyard, which contained two large stacks of baled alfalfa hay. They were nearly out of diesel fuel. They parked the Cat at ranch headquarters and got into their pickup. The driver told me they were going to town for diesel but would be back at the ranch before eight in the morning. They couldn't plow any more roads for us, but had to open the road on south past other ranches.

With the road now open up Skull Valley, Walter Dansie arrived about five o'clock in the afternoon, just after the Caterpillar operators had left. He, Will, and I had a conference. We had a real emergency. Mr. Dansie said he had called Fife, our friend in Evanston, Wyoming, who did the reservoir-building and road work for us on the summer range. We could get two of his D7 Cats, and a driver for each, to come here to Skull Valley in a week. Dansie continued that he had purchased a new big truck that would be here the next day, so we would be able to haul more feed. The BLM had made arrangements to have a Cat transported to Low, over on U.S. Highway 40. It was supposed to get there in the morning and plow the road to the north on the Lakeside grazing unit. We could use it the following day, but we needed to meet the Cat operators the next morning at Low and make arrangements for them to help us. I said I would be at Low early in the morning and talk to them about helping us the next day. We could have our two big trucks, loaded with feed, follow the Cat to the west side of Cedar Mountain the day after next.

Just then Glen Hess came up and told us we could help ourselves to hay in that first big stackyard where the Cat opened the road. The ranch had plenty of baled hay, and Glen would have the teams feeding the cattle go to those stackyards where we couldn't drive the trucks. Things were working out rather well. If the two Cats came from Evanston as anticipated, Will would use one in Skull Valley, and I would use the other on the west side of Cedar Mountain. Mr. Dansie said that if a miracle occurred during the next few days, and the snow melted, we could cancel the Cats from Evanston. We concluded the conference, and he returned to Salt Lake City.

The next morning, I was at Low at seven o'clock and arranged for the Cat to help us all the following day. When the time came for the Cat to work, the two big trucks loaded with hay were there, plus my Jeep, which carried three 50-gallon barrels of diesel fuel for the Cat. As we left Low and headed west on Highway 40 for a couple of miles, the Jeep took the lead until we turned south off the highway. Then I pointed out to the Cat driver the road that needed to be cleared—down south across the railroad tracks at Aragonite, then east to the foot of Cedar Mountain. The Cat driver was to stop and wave to me when he couldn't see the road. The Jeep followed just behind the Cat.

We reached the foot of Cedar Mountain and proceeded south past two herds of sheep. Although the sheep were in deep snow, they were still able to graze sagebrush and juniper trees. The herders and camptenders had their teams hooked up to large junipers that they had cut and were now dragging through the snow to make trails for the sheep to follow. This was old-time standard practice. We unloaded the hay at the stackyard where the north rider camped, then had lunch at the camp. The two big trucks returned to Skull Valley, and the Cat driver and I worked on south past two more herds of sheep. The men were very apprehensive about the situation, but seeing the Cat and talking to me helped give them a sense of security, at least for the time being. There were over twenty inches of snow, which could be real trouble. I told the herders and camptenders to freely use the hay in the stackyard and the pellets in the granary, and that I would be back in a few days.

The snow subsided for a few days and settled down three or four inches. The road up Skull Valley was kept open, and I was still headquartering at the ranch. Will Sorensen and I worked together with the two big trucks and drivers as we hauled feed out to the sheep in Skull Valley. Usually one truck was loaded with baled alfalfa hay, and the other loaded with 100-pound bags of grain pellets. These trucks were constantly chained up, in order to pull through the deep and drifted snow. Now the objective was to get feed to a different herd of sheep each day.

One morning, I prepared to leave the ranch in the Jeep pickup, in order to visit the herds on the west side of Cedar Mountain. The two big trucks, loaded with hay, were to meet me about noon the next day at Clive on Highway 40, ten miles west of Low. I told our men I planned to visit the four herds close to Cedar Mountain, then go on south and west and stay overnight at the camp at Rattlesnake Point. The next morning, I would travel north to Clive, a distance of twenty-five miles, and meet the trucks as arranged. We would then transport the hay to Rattlesnake Point, if that was the decision after looking over the situation. That night we got more snow at Rattlesnake Point. I left the camp about 8:00 a.m., and headed north to meet the trucks at Clive. The pickup was constantly in four-wheel-drive, the front bumper pushing snow, my progress about eight miles per hour. Going through a drift thirty inches deep, the engine sputtered to a stop. The problem was evident when I raised the hood—the fan had thrown snow over the engine. The heat from the engine melted the snow, and water around the sparkplugs caused them to short out. I used paper from a roll in the emergency box and dried the sparkplugs. The engine started, and we were off again. In just a few minutes, the sparkplugs shorted out when the fan threw snow over the engine again. Here's where Grandad's big overcoat solved a problem. I carefully put the coat over the engine, so the sparkplugs were covered but the coat wouldn't catch on fire from the hot engine. No more shorting occurred.

The Union Pacific railroad track was one mile south of Highway 40 at Clive. A small railroad section crew lived at the railroad siding, close to where the Jeep had to cross the railroad tracks. As the Jeep approached the railroad crossing, two railroad men came out to wave hello. The two big truckloads of hay were parked at the Clive service station and café on the highway. I made it at noon. The men were amazed when I lifted the hood and removed the overcoat. Each of us ate two hamburgers at the Clive café as we talked over strategy. The wind was not blowing, and the sun was shining brightly. I said we should go on to Rattlesnake Point with the hay. The trucks were already chained up, and we filled them with gasoline at the service station. Although progress was slow, we made it. I told Candido and Cornelius to use the hay, and that I would see them in a few days. It was after dark when we got back to Highway 40. We ate hamburgers at Clive and returned to the ranch in Skull Valley at nine o'clock.

14

APPROACHING DISASTER

THE SNOW CONTINUED TO FALL. The two D7 Caterpillars and drivers arrived from Evanston. One was transported to the ranch in Skull Valley, and one went directly to the west side of Cedar Mountain. For the immediate future, Will Sorensen would take primary responsibility for the herds of sheep in Skull Valley, and I would work to get feed to all seven herds on the west side of Cedar Mountain. The day after the Cats arrived, there was another heavy snowfall. Returning from the west side of Cedar Mountain, I found the road up Skull Valley closed, absolutely blocked with heavy snow. The two big trucks were at the ranch in Skull Valley. This cut off my supply of feed. I had the Jeep pickup truck, one D7 Cat, and Red, the driver, on the west side of Cedar. I called Mr. Dansie to explain the situation.

In about an hour, he returned the call. He could get baled hay, which was available in Kansas. He said the Union Pacific Railroad would deliver five carloads of baled hay to us either in two days, or three. He needed to know where I wanted those cars. I told him we needed two cars set on the siding at Aragonite, just west of Low near Cedar Mountain, and the other three put on the siding at the Clive station, way out west. Mr. Dansie told me to purchase grain pellets at the feed mill in Grantsville and to hire two men with their trucks. Two locals agreed to haul pellets from Grantsville and then haul the baled hay from the railroad sidings.

In order to keep the Cat working, it had to be kept in diesel fuel. Three 50-gallon barrels of diesel in the back of my pickup were the source of fuel. These barrels had to be refilled in town. Home for the night in Grantsville, it was a wonderful reunion with Kathryn and our four small children. I loved each one for a few minutes, then shaved, bathed, and put on clean clothes. Then, more loving, as Kathryn and the children told me about what was going on in Grantsville. I emptied my suitcase full of underwear, shirts, and socks to be laundered, then repacked it with clean clothes to last a few days back out on the desert. I told Kathryn two trucks loaded with grain pellets were ready to leave town at six thirty in the morning. They would follow me to the west side of Cedar Mountain, where the Cat would clear the road for us to get to the herds of sheep. I needed to eat breakfast at six o'clock.

Next morning, the two trucks followed me, and we fell in behind a procession of six big commercial trucks heading west out of Grantsville. There was a foot of snow on the highway. Those heavily loaded big rigs broke a trail. It was a good day. Red drove the Cat, and we got the feed to where we wanted it. The two truckers returned to Grantsville and agreed to bring another two loads of pellets the next day. The following day was a rough one on the desert. It was cold, with a stiff wind blowing. The two loads of pellets were unloaded, and the men returned to Grantsville. In the late afternoon, Red and I headed south to the herd at Cedar Spring, south of Rattlesnake Point. We made it just after dark. The herder and camptender, Joe Lucero and Adolfo Salazar, heard the Cat coming and had plenty of dinner for all four of us. Joe and Adolfo took the bed for the first half of the night, from eight o'clock until two thirty, then got up and dressed, so Red and I could stretch out and sleep in the bed until morning.

After breakfast, when we stepped out of the camp, it was a shocking scene. The hair was gone from the horses' tails. Joe saw me staring at the horses and said the sheep had eaten the hair. When we walked out into the sheep, wool was gone from the backs of some of the rams. Joe told me some of the rams had a few hay leaves in the wool on their backs when they came to the herd in December. The sheep had eaten the wool in order to get the hay leaves. I told them that we would get feed to them tomorrow or the next day. We needed to go on over to Luis's camp, close to Wig Mountain, but didn't have enough diesel fuel. We had only enough diesel to get the Cat back to the highway at Clive, so we'd have to get more fuel there and come back the following day.

Red and I made it to Clive, where we left the Cat for the night. Red wanted to go to Grantsville, where he could get a motel room and take a shower. When we got to Timpie, we stopped at Rowberry's service station, and were told the road up Skull Valley was opened up again that day. Red said he would eat dinner at Timpie, while I went up to the ranch to see Will Sorensen. Traveling south up Skull Valley, the snow was piled up four or five feet high at the side of the road. In two places, the snowplows had cut through drifts over eight feet deep. It was like driving through a small canyon, with vertical walls of packed snow.

Snowdrifts higher than motor vehicles on the Skull Valley road.

At the ranch, Will Sorensen and I had a conference. He asked if I had been able to get to Luis's camp at Wig Mountain. I told him we had used all of our diesel fuel getting to Cedar Spring and back to Clive. I was on my way to town for diesel and could get to Luis's tomorrow. I had two truckers from Grantsville hauling pellets, and I needed to have them follow us tomorrow. Will said that he and the men had been to Paul Silva's camp at White Rock that day, and that Luis's camptender had been there for three days. His horse fell into a snow-covered ravine on his way to Paul's camp, and they had a hard time getting out. Both the man and the horse got hurt, but there were no broken bones. The horse was lame but getting better. The camptender figured he could go back to Luis's camp the day after tomorrow, but Will told him I probably had already been there, and for him not to worry. I told Will that it was taking more diesel than I figured, but we'd been to all the camps except Luis's. We'd get there the next day, as it was thirty miles from Clive to Wig Mountain.

I asked Will how he was getting along here in Skull Valley. He replied that it was a bad situation. This was the worst he had ever seen it. The Cat plowed out a road, then the snow drifted in and filled it up, and then the snow packed harder than ever. They had three or four roads, side by side, going to White Rock that had just drifted over. It was easier for the Cat to break a new trail

South rider Paul Silva and bay horse John, March 1949, near his
camp at White Rock.

than plow out the old one. Will was not happy. I headed back to Timpie. Red
had eaten dinner, and we drove to Grantsville. The snow had been bulldozed
to each side of Main Street, which was also U.S. Highway 40. Grantsville had
two parallel drifts of snow, eight feet high, running all the way through town.
An entry had been cut to each side street, and to some of the places of business. We got Red a room for the night. We'd leave at seven in the morning. I
got four barrels of diesel; the truckers were already loaded. Then I went home
to Kathryn and the kids.

Morning came early, and we left town right on schedule. We unloaded
one truck at the camp at Cedar Spring and took one on south another seven
miles to Luis's camp near Wig Mountain. Luis heard the Cat coming and
came to the door of his wagon. He was shading his eyes with his hands, as he
had gone "snowblind," and could hardly see. He had fallen several times in
the deep snow and had lost his dark glasses. Luis said his camptender left on
his horse five days ago, headed for Paul's camp at White Rock, and he hadn't
heard what happened to him. I informed Luis of the situation, saying the
camptender was at Paul's camp, but coming back tomorrow or the next day.
Luis said he needed some wood, as he was down to the last piece. Red took the
Cat to drag in a dead juniper.

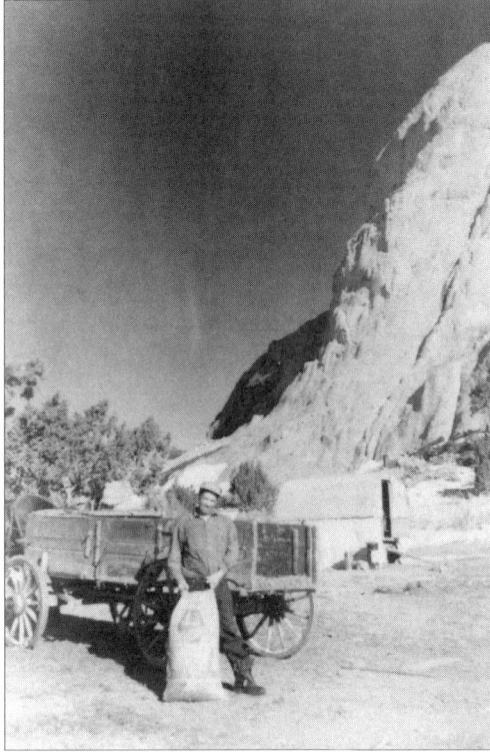

Author's Collection
Dan Gallegos at Paul Silva's camp at
monumental White Rock, in south
Skull Valley, March 1949.

The sheep were widely scattered up over the hills to the east. None were
very close to camp. I said to our men, "Don't unload these pellets." I walked
out about a mile to look over the situation. The sheep were spread out through
the hills for a couple of miles. Under the circumstances, they should be left
alone. They had gravitated to the southern slopes, where there was slightly
less snow. The days were getting longer, and the snow on the south slopes was
settling down.

Back at camp, I gave Luis my dark glasses, saying I could buy glasses over
at the highway. I told Luis that his sheep looked to me to be in the best possible situation. We didn't need to disturb them, just let them work their own
way up the hills. I said I'd be back in a few days, and asked about his food
supply. He replied that he had lots of food. By then, Red had cut a big pile of
wood. We all shook hands, and Luis was alone again as the rest of us headed
back north to unload the truck at Cedar Spring.

Standing, from left, Dan Gallegos, Tim Wilmer; seated left,
Paul Silva; right, unidentified, March 1949.

Things continued to move well. I was able to keep in close touch with Will Sorensen and with Mr. Dansie. Finally, it was an opportune time to move Candido's herd from the area at Rattlesnake Point to the foothills of Cedar Mountain, five miles east. We agreed on the strategy. Red drove the Cat and cleared a pathway the width of the dozer blade through the snow. We started the sheep out on to the cleared trail, and they followed along behind the Cat. The whole herd of over 2,500 ewes was strung out in a long trail for two miles. The snow was two feet deep on each side of the cleared path. Then a breeze started to blow from the south. As the wind gathered speed, the snow started moving horizontally, at ground level. By now Red was to the mountain, and the sheep's leaders were right behind him. He cleared a road to a big south slope, and over a thousand sheep were able to make it through.

Candido and Cornelius were on their horses, about a mile from the mountain, using their dogs to help move the sheep. I was walking at the rear of the herd, pushing the sheep. When it was obvious the trail was drifting over and covering the sheep, I changed tactics and let several hundred sheep run back to the old campground. Trudging on foot through the snow, I made it up the trail about a mile and a half and came upon little bunches of five to ten sheep, huddled together and being drifted over. I turned them back to the west and gave them a shove. Some made it back.

Red was heading west again with the Cat, and plowing out little bunches of sheep that were caught in the drifts. Candido and Cornelius were pushing sheep into the newly cleared trail Red made, and were successful in getting another couple of hundred to the mountain. Now the wind became a real horizontal blizzard. I couldn't see thirty yards in any direction. The sheep I stumbled into just wouldn't move. It was time for me to turn around and head back west toward my Jeep. As I struggled that mile and a half and got back to the Jeep, I knew we were going to lose several hundred sheep this day. Somehow, the plight of the Marines left on Wake Island, as they faced up to superior Japanese forces landing on the island, was running through my mind. The last communication from the Marines was, "The issue is in doubt." So was today's objective. As I contemplated several hundred frozen sheep covered with snow, I knew we were dealing with animals, not humans. This was a rough day at Rattlesnake.

That afternoon the wind subsided, and we followed Red as he plowed snow from the original Cat trail where the sheep had been drifted over. The sheep we found were dead, frozen stiff. Candido figured 1,200 sheep had made it to the mountain. Just before dark, we fed the eight hundred that made it back to the old camp. We knew five hundred were dead in the drifts. It was now dark, and all four of us were just simply half-frozen and worn out. None of us had ever seen such a one-day disaster. During supper, we discussed moving the remaining eight hundred ewes to the mountain in a few days, maybe tomorrow if things looked right. We were grateful that two thousand sheep were still alive in that herd. It was a long night. As usual, the herder and camptender slept in the bed from about eight o'clock until two o'clock, then Red and I had the bed from two o'clock until six. Next morning, the sun came out and things again looked good. Red cleared a new path, and we moved the remaining eight hundred sheep to the mountain, where they joined those that made it though the day before. Cornelius hooked up the team and moved the camp, using the path Red had cleared.

East-west U.S. Highway 40 became drifted in and closed for a day or two several times that winter. Eventually, the snow along Main Street in Grantsville was piled so high it almost reached the utility wires. School children were constantly warned to not climb those mountains of snow. It is possible to touch a live wire and be electrocuted.

In early March, as the days were lengthening, the snow began to melt. In Skull Valley, alongside the road, there were still heavy drifts several feet deep. Will Sorensen and I were at the ranch, talking over the situation, when Will said that this was by far the worst winter he had ever seen. He set our winter loss at ten thousand sheep. I replied that such a loss seemed a little heavy to me. Some sheepmen had predicted a 20 percent average loss for the winter throughout the western range country. Our heaviest losses would be on the south end, on the west side. Will repeated that he had set this loss at ten thousand head and said if it turned out to be less than ten thousand, I'd be a mighty lucky young man.

Looking Up

On the west side of Cedar Mountain, at the edge of the flat area at the mouth of Quincy Canyon, lay the rusted-out shell of what was once a 2,000-gallon horse-watering trough. It had been installed in the early 1900s, when the Standard Horse and Mule Association grazed several hundred horses year-round in this area. This horse association had their last roundup in the late 1920s. Originally, they piped water down to the flat from Quincy Spring, about a half mile up the side of Cedar Mountain. The underground pipeline had long since gone the way of the water trough and was no longer functional. I considered building a sheep-watering facility to take advantage of Quincy Spring, but figured a large, open, flat area three hundred yards up the canyon would be the logical place. Red cut a road, cleared away sagebrush, and leveled the earth to eventually accommodate a 3,000-gallon storage tank and 150 linear feet of metal water troughs. The displaced earth needed to settle for a year. I told Will that the next year, or the year following, we could install the tank and troughs and pipe the water, to make another watering place.

On March 21, 1949, the annual stockholders' meeting of the Deseret Live Stock Company was held at the Newhouse Hotel in Salt Lake City. At that meeting, Mr. Dansie reported that we took our "40,000 plus" sheep to the desert in the fall of 1948. We had a terrible winter of deep snow, and our loss was 15 percent. He reported that the net profit for 1948 was $139,000, and federal income tax amounted to $84,000. A 6 percent dividend for 1948 was paid to all stockholders.

That spring of 1949, Riley Hale and his brother Tom, both from Grantsville, were hired to skin the pelts off company sheep that had died out on the range. They skinned in excess of 1,300. When the pelting operation was finished, both Riley and Tom wanted to keep working for the company.

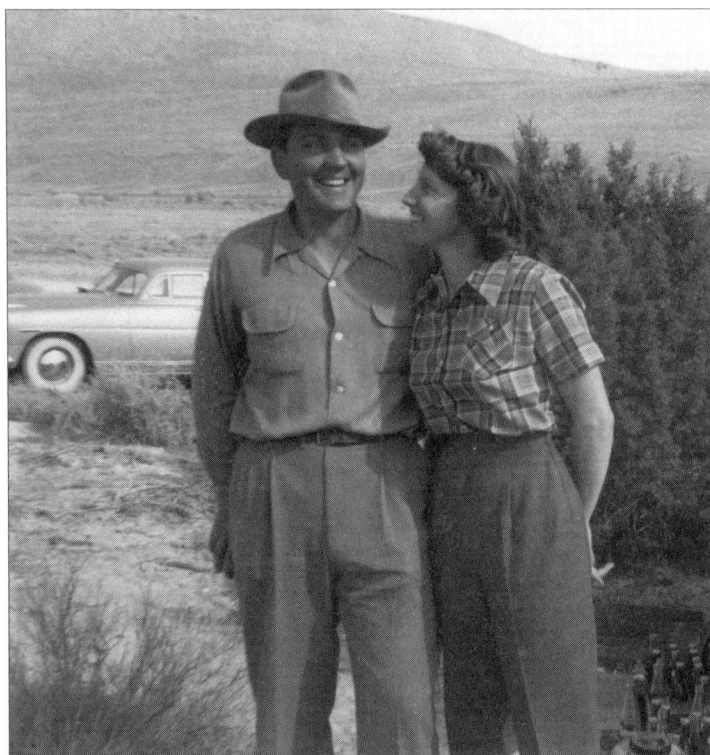

Author's Collection

Easter, spring of 1949, Dean and Kathryn Frischknecht at
Stansbury Island in the Great Salt Lake, where friend Jess
Charles is preparing a Dutch-oven dinner for several friends.

They were experienced in many areas of livestock production and took over
a herd of sheep.

During the late winter, I suggested to Will Sorensen and Mr. Dansie the
desirability of "tagging," sometimes called "crutching," the lambing ewes
sometime in April, while they were still on the winter range. Tagging is the
process of shearing the wool away from the crotch and the udder, although the
rest of the fleece is left intact on the sheep. This makes it easier for newborn
lambs to find the teat and nurse. It would also keep the crotch area cleaner,
and prevent the problem of flies infesting that area when it became wet with
manure and urine. Because lambing took place at high elevations in May, the
Deseret Live Stock Company did not shear until June, after lambing.

To facilitate tagging, we built a corral near the well on the north end of
the west side of Cedar Mountain, another corral near the well on the south
end in Skull Valley, and a third corral at the ranch. Nels Frandsen and his crew

were hired to do the job. The herders were pleased. The removal of a third of a pound of wool from each ewe would save lambs. The thirty large bags of wool we sheared then averaged nearly four hundred pounds per bag. The wool was trucked to the railroad siding at Timpie, where it would lie unprotected overnight. I told Mr. Dansie and Will that I would take a little bedding and spend the night in my car, sleeping near the wool. They both agreed that it was necessary.

Finally the snow melted; spring became beautiful. Plants responded to the excess moisture and warm temperature. The range was a verdant green. When spring arrived, Mr. Youngberg and his crew from Salt Lake City began building a new and larger main house at the ranch in Skull Valley, on the exact location where the old house stood. This new house would have the great advantages of a big new kitchen and a larger dining room.

Because of snowdrifts and a late spring at Wahsatch, the BLM allowed the sheep to stay on the desert for an extra week. Shipping was done May 4, 5, 6, 7, and 8; Will went early to receive them. By now feed was abundant on the desert, and the sheep were in good shape. Now that winter and spring losses were determined, Mr. Dansie felt relatively good. After the fourth day of shipping, with only one day left, he told me that this was not nearly the greatest loss of sheep ever experienced by the Deseret Live Stock Company. Several years ago, they lost over sixteen thousand sheep in one bad winter and spring. At that time, the sheep had to walk from the winter range to the summer range.

That spring, in May 1949, there was still a lot of snow on the north sides of the hills at Wahsatch. The country warmed up fast, and soon there was an abundance of feed. There were fewer lambing herds than usual, and the 5,500 surviving yearling ewes were put into two herds of 2,750 each. Emma Zabriskie came back to Wahsatch and cooked from the first week in May until shearing was finished in July. We appreciated her. Our son "Bill" Dean (who is officially on the school records as "Dean") finished first grade in Grantsville in May. I then transported Kathryn and our four children to our home at the shearing corral.

The board of directors called a special meeting of stockholders for June 7, 1949, at the Newhouse Hotel in Salt Lake City, to discuss the acquisition of land in Florida. At that time, the general consensus was for the company to continue exploring the situation, but not to purchase the land yet.

As shearing time approached in 1949, Mr. Dansie hired a shearing crew from Texas. He hated to quit using the shearing crew from Utah, but the Gonzales crew, Spanish-Americans from Texas, did the job considerably cheaper. This eliminated the big job of our company doing extra cooking for the shearers. We had to furnish mutton or other meat for the Texans, but their manager would furnish the balance of their food. Their cooks would feed the men by their tents, close to the shearing shed.

Shearing got underway June 15, 1949. The rams were shorn first. I examined each foot on every ram for foot rot but did not find any. Mr. Dansie had told me that when the yearling ewes were shorn, he wanted me, or one of our men who could recognize foot rot, on the shearing floor. We had a few of the replacement-ewe lambs brought into the ranch during the winter, and they had been in a corral where the Merino rams were previously. If we found a yearling ewe with foot rot, it had to be slaughtered, right there, that evening. We couldn't take the chance of foot rot getting started in our ewes. However, we knew the yearlings were in good flesh, and their meat was edible. Paul Silva, Ace John, Bill Watts, and I organized it so two of us were on the floor at all times. We examined each foot on each yearling ewe. We found four ewes with a light case just beginning, so the slaughter assignment was carried out on them. That ended the foot-rot episode; no ewes got the disease.

Tagging the lambing ewes was a big help, but we needed to make some changes. In the coming fall, we needed to vaccinate our replacement-ewe lambs for "soremouth" (contagious ecthyma). That disease caused some losses in the ewe lambs on the desert. I had counted each herd, and told Mr. Dansie and Will that we lost three thousand of the replacement-ewe lambs. They couldn't handle the deep snow. We needed to have them where they could be fed, if necessary. Will and I thought we should build two additional stackyards and granaries on the winter range, where we could stockpile emergency baled hay and grain pellets.

At the shearing corral one day, Mr. Dansie asked me how many lambs we docked that spring. I told him over twenty-two thousand. He replied that that was as many as we docked on lots of good years before I got there. I told Mr. Dansie it was a good thing Will Sorensen had still been on the job last February. He had looked after the sheep in Skull Valley, while I was getting feed to the seven herds west of Cedar Mountain.

Ace John and his wife had a small son, age six, and it was going to be a difficult situation for Ace to be away from his family as a rider during most of the year. He was offered better employment at the new military base at Dugway, at the south end of Cedar Mountain. It was best for him to leave the Deseret Live Stock Company, so I would again handle that north rider job for the summer.

It was a glorious summer. Several days each week, Kathryn and the kids rode with me in the pickup as we visited different camps. Bill, Diane, and Dale preferred to ride in the back of the truck, but Doris, now over seven

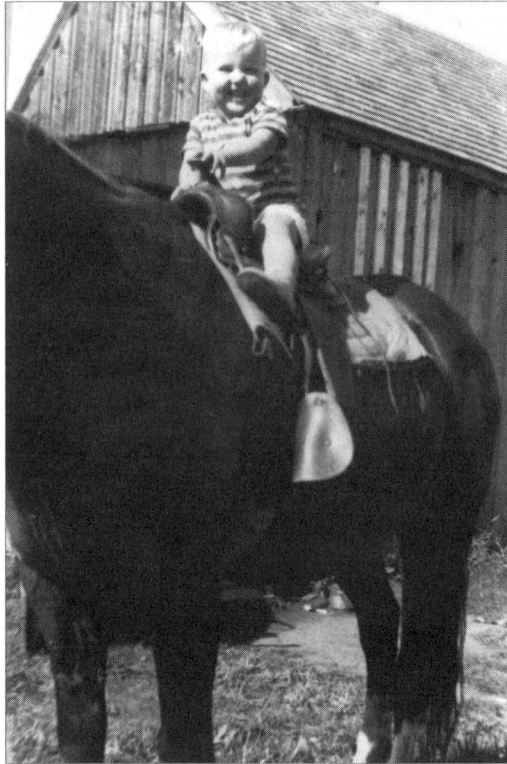

Author's Collection
Dale Frischknecht on Old Bally at the
Deseret Live Stock Company's shearing
corral, May 1949. Our two-room house is in
the background.

months old, was usually in Kathryn's arms, or sleeping on the seat. As usual, we attended the summer picnic of the Monte Cristo Grazing Association. Although most of the Deseret Live Stock Company's cattle and sheep grazed on private land, the company belonged to the grazing association because of the two herds of sheep that grazed on the Cache National Forest. The picnic was held at a public park close to Monte Cristo Peak in the national forest.

Will and Vera Sorensen took off for a couple of weeks in their new Cadillac and stayed in their home at Spring City. Soon the word reached me that Will told people he now had to train a new sheep foreman. That was bad news to me.

A Great Recovery

WHEN MR. DANSIE CAME TO the summer range a couple of weeks later, I pulled him aside for a conference. I told him word had come to me that Will Sorensen blabbed around Spring City that he really wanted to retire, but couldn't, because now he had to train another new sheep foreman. I said, "If you fellows plan to let me go on account of last winter's sheep losses, I need to know." Mr. Dansie was totally surprised that Will had talked and that now I knew. He said after that bad winter of deep snow and poor feed, the board of directors figured they had better hire another man to help run the sheep, since it was a good thing Will was there and still able to help last winter. He stated that Will did want to retire, but the company wanted him to stay and help train someone who could help manage the sheep. After all, forty thousand sheep on the winter range could be a difficult situation for one man to manage. He concluded by telling me, "We have a big investment in you. We have no thought of replacing you. We want to help you." This lightened the load Kathryn and I were carrying, but I was not about to discuss it with Will. He had chosen to leave me in the dark.

Kathryn and the four children left the mountain in late August. Young Bill would be in second grade, and Diane would start first grade. Both were reading well. The fence crew finished work the last of August, and the high-school-age boys returned to school. Bill Watts moved over to summer head-quarters to help with the fall work. In the late summer of 1949, William D. "Bill" Cook came to work for the company. Mr. Dansie, Will, and I felt it would work best to put Bill Cook in charge of the south half of the outfit, and have me in charge of the north half. Cook was from a well-known sheep-raising family in Fountain Green, Utah, and was experienced in sheep operations. His mother was a Livingston from Manti, a sister to the previously mentioned

Joe Livingston. Will told me to turn "the books" over to Bill Cook, and show him how to handle that responsibility.

In early September, Will Sorensen was trying to burn some large patches of sagebrush on the lower range near Peck Canyon, around the south end of Salaratus. He set several fires, but didn't get much accomplished. After a hard day with little success, he gave up in disgust and came back up the mountain to summer headquarters. He told me that he would get a fire going pretty good, and it would burn a small patch of sage. Then it would die out because there wasn't enough breeze to carry the fire to the next thick patch.

Five days later it was lamb-shipping time, and we were out on the south end, at the corral at the head of Horse Ridge, separating the sale lambs from the ewe herds. Along in the late afternoon, a forest ranger with an urgent message drove up in a pickup. He said that down on the lower range, there was a bad fire raging north through the Salaratus area toward Home Ranch headquarters. The forest service already had a large firefighting crew on the front edge of the fire, which could burn fields and haystacks, and maybe houses, through the Home Ranch and on into Woodruff. This fire started in Peck Canyon, from where Will had ignited several fires a few days ago. Will was surprised and insisted the fires he started had died out. Will said that some tourist, or else lightning, must have started this fire. The ranger concluded by telling us that Mr. Dansie wanted us to know about the fire, as well as wanting Will to go take a look. I thanked the ranger for coming.

Will told me he had better go see about this fire. I agreed and said that Bill Watts should go with him, in the little Jeep. I added that they had better take some food and water from my house. Will and Watts left immediately. They came back to headquarters during the night, about 2:00 a.m. The next morning, Will was smiling and said that the burn cleared away the brush on five thousand acres. It was a beautiful burn, and the fire was put out by the time they got there. Will said he wouldn't have minded having credit for that burn, but it couldn't have originated from any of the fires he started. This fire hadn't begun until five days after he was down in Peck. Because of the fire, we'd have a big job of reseeding to do in the fall. He added that he hoped Walt Dansie could get up here to see it in a couple of days.

The south-end lambs were trailed to Wahsatch, then trucked to Ogden. The north lambs were trucked from the summer range, going directly off the mountain and west to Ogden, where Mr. Dansie received them. My note to him said that we should be done cutting out the lambs by noon on September 27. If the trucks made one trip on the twenty-seventh, we ought to be all cleaned up by then. I believed we could have the old ewes at the shearing corral at Wahsatch on the twenty-ninth, and possibly ship culls on the thirtieth.

This we did. That fall, we again culled sheep as usual, but didn't have many culls. Winter had largely taken care of the culling. The surviving sheep, however, were strong and in excellent condition. During October 1949, we again dipped all of the sheep at the dipping vat in Salaratus. Things went well. Only a few sheep were infested with ticks, and dipping completely eradicated all external parasites.

Paul Silva had his wife, Susie, living with him during the summer. He decided that now was a convenient time to terminate his employment. He told me we had had good times and one bad winter. He had enjoyed working with me, but now there was plenty of new help, so he and his wife were going home to Ignacio, Colorado. Paul had done a good job as the south rider. We shook hands, and I told him that I appreciated all we had gone through together. If he ever decided to come back, he would have a job.

Bill Cook arranged for his younger brother, Grant, to help as the rider on the south end. Grant was young and husky, a good worker. Will Sorensen, Bill Watts, and Bill and Grant Cook took over the responsibility for reseeding the five thousand burned acres. It took several days, and they did a thorough job. While they reseeded, I worked on the sheep over at the shearing corral. We kept 7,600 replacement-ewe lambs, so the January 1, 1950, inventory of sheep would be within range of the inventory a year earlier, before the heavy losses due to deep snow.

Much of the sheep loss in the winter and spring of 1949 was due to soremouth in the replacement-ewe lambs. Therefore, in the fall of 1949, we vaccinated all replacement-ewe lambs in order to prevent this disease. I had told Will that we would have to put this vaccine on bare skin, in places where there was no wool. He replied that rather than catch each ewe lamb and vaccinate it in the flank or under the tail stub, he thought we could do it on the inside of the ear as they stood in the branding chute. That would save a lot of hard work. It was a great idea.

Vaccinating in the ear eliminated the strenuous job of setting each animal on its rump in order to vaccinate it in an area free from wool. When the ewe lambs were crowded tightly into the mouthing chute, four men easily did the job, working from outside the chute, two on each side. The first man extended one arm under the sheep's neck, and grasped one ear between his thumb and forefinger. Then, with a knife in his other hand, he lightly scraped the exposed skin on the inside of the ear. The second man carried a small bottle of vaccine and a tiny brush for applying a drop of vaccine to the scraped area. The application was made while the first man held the ear steady.

Mr. Dansie came to the shearing corral and explained that he would like to send two thousand ewe lambs to the Imperial Valley in California, to winter

Bill Watts at the Skull Valley Ranch, December 1949. This truck could be made into a "double-decker" when needed for hauling sheep. Bill worked for the Deseret Live Stock Company before and after his World War II service.

on the alfalfa fields there, at a cost of five cents per head per day. The total cost would be ten dollars per head from now until about the tenth of May, including freight on the railroad both ways. Mr. Dansie thought we should cut out and send the smaller ewe lambs. We got the job done, and it was obvious that this crop of replacement-ewe lambs would be well taken care of. The remaining 5,500 were divided into three bunches of about 1,800 each, then 700 older ewes were added to each bunch, to make three herds of 2,500 sheep each. These herds would winter on the north end, near Eight-Mile Spring on the east side of Cedar Mountain, where it would be convenient to feed them supplemental protein and energy pellets. We wanted to teach these ewe lambs to eat pellets as a supplemental feed, and the extra nutrition would help them continue to grow throughout the winter.

Sophus Christensen came as usual, and eye shearing was completed. The summer wagons were stored in the warehouse, and the extra horses—the ones that were not going to be used during the winter—were driven to Home Ranch headquarters. Shipping to the winter range was done in five days, as it usually was, in early November. Things were going well. Feed on the desert was plentiful. I had previously asked Bill Watts to be the north rider, but he was not enthusiastic. However, now he agreed to do the job for the winter. He knew the range, and was a single man, so he could stay out on it.

Keith Moss McMurrin Collection
Deseret Live Stock cattle on the move.

We had just shipped the sheep from Wahsatch to Skull Valley that fall of 1949 when word came that Ralph Moss, foreman at the Home Ranch near Woodruff, Utah, had suffered a fatal heart attack. I was particularly sorry to learn about this. I told Kathryn that all of my memories of Ralph Moss were positive. Many a time he had given me fresh young saddle horses and teams of young draft horses. He was good in sending us a quarter of beef when he figured we could handle it. Ralph knew I was just as much interested in the cattle and horses as I was in the sheep. We always talked about how the cattle operation was managed. He was good at explaining things to me. Ralph and his wife had raised a fine family, and Ralph would be sorely missed. Ralph was a son-in-law of B. H. Roberts, who had been a very prominent leader in church and civic responsibilities in Utah. Peter Mower, originally from Fairview, Utah—another 'Sanpeter"—and a long-time employee at the Home Ranch—replaced Ralph Moss as foreman over the cattle operation. He had many years' experience working under Ralph. Pete did an excellent job, and things continued to go well at the Home Ranch.

When the herds were on their ranges, we butchered three market-ready cattle and distributed meat to all the camps. Bill and Grant Cook headquartered in a camp at White Rock, but helped build a new metal granary, a stackyard for baled hay near the Eight-Mile Spring on the east side of Cedar Mountain, and a similar granary and stackyard out south. Our crew trucked an emergency supply of grain pellets and baled hay to these new granaries and

Pete Mower, Deseret Live Stock cattle foreman, atop well-used tractor.

stackyards, and another set to the north rider's camp, on the west side of Cedar Mountain. Mr. Dansie purchased 675 tons of grain pellets in 100-pound bags, for a total of 13,500 bags. A lot of feeding was done that winter. The old ewes and replacement-ewe lambs wintered near Eight-Mile and were fed pellets every day. They really "did well."

The routine job of hog killing and pork processing was accomplished in early December. The men at the camps enjoyed the fresh pork, particularly our homemade sausage with some ground beef included. Rams were branded and trucked to all the ewe herds on the regular days, December 15 and 16. A few thin ewes were hauled out on the return trips to the ranch. Mr. Dansie came to Skull Valley with the Christmas checks a week before Christmas, and most of the men once more mailed these checks home.

I told the men that Kathryn and I would be in our home in Grantsville for Christmas, and then we were going to visit her folks and mine in Manti for a couple of days. That Christmas was another great one, and after it was over we put all four kids in the front of the Jeep pickup and went to Manti. The seating had to be just right. While in Manti, I looked at new Hudson sedans at Nell's Motor. A short time later, Art Nell made me a very good price on a new gray Hudson, so we bought it. This gave Kathryn and the children dependable transportation while I was gone out on the range.

17

ALL IN THE DAY'S WORK

ONE DAY, GLEN HESS TOLD me there were more than three hundred wild horses within twenty miles of the ranch. A couple of fellows who were experienced in rounding up wild mustangs were coming the next day and would spend two or three days gathering horses. They'd have a Piper Cub airplane to maneuver the wild bands in close to the ranch. Then Glen and the cowboys would get them running west, down the outside of the two-mile-long north pasture fence. Glen went on to tell me that when the mustangs rounded the corner of that long fence, they'd run right into a high, woven-wire corral. They'd be inside the corral, and caught, before they knew it. The fence was strong and too high for them to jump.

Glen and the cowboys worked on this horse roundup. After a wild bunch was inside the first corral, they were turned into a fenced-in pasture, along with a dozen gentle horses. Mounted cowboys, quietly working at a distance, moved the horses around another fence into a high-pole corral. There were 150 wild horses in our corrals. Some owners came to claim a branded horse. Sixty genuine but poor-quality mustangs were culled out and taken off the range. Some of the better-looking yearlings were for sale. The best-quality two- and three-year-olds were loaded up to go to a ranch and be domesticated. Two big stallions and twenty-five of the best mares were turned back on the range.

In early April, I was ready to complete the water project at Quincy Spring, on the west side of Cedar Mountain. We hauled 1,800 feet of galvanized pipe, and strung it from the leveled area to the spring, 1,800 feet up the mountainside. We dug a long trench, connected and laid the pipe, and piped water to the 3,000-gallon metal storage tank. An overflow pipe at the top of the tank allowed the water to flow into 150 linear feet of newly installed troughs. This was a great improvement, because it put the water to an accessible open, flat area. The full troughs held 750 gallons. The herder had to open the valve at the bottom of the storage tank in order to release its 3,000 gallons into the troughs. This was plenty of water to accommodate a large herd of over 2,800 ewes. When the sheep finished drinking, the herder closed the valve so the storage tank would refill.

At the annual stockholders' meeting, held April 10, 1950, in Salt Lake City, Mr. Dansie reported that the Deseret Live Stock Company had not purchased the land in Florida. President Henry D. Moyle indicated if the company did not wish to purchase the land, he would proceed in another direction in order to still acquire the property. He thought it was an opportunity that should not be passed up. He was given the green light to go ahead. President Moyle arranged for and consummated the purchase of the Florida property. The options to purchase it were in the name of the Deseret Live Stock Company. Since the company in Utah did not purchase the land, the ranch in Florida was purchased and developed under the name Deseret Farms of Florida.

I was able to go home to Grantsville overnight, two or three times each week. On Sundays, we often were able to attend church as a family, and it was uplifting and positively helpful. Green grass was springing up all over the range in April 1950. The Frandsen shearing crew once again came to the desert and tagged all the lambing ewes.

One day Nels Frandsen told me that I worked too hard. He saw me lift ten or twelve ewes over the fence, because we could see they were not going to lamb that year and didn't need to be tagged. That saved the company a few cents on each one. Nels said that I went all day, like a machine, and one of those times my ticker was going to quit. He was right. Mr. Dansie had told me to slow down on the physical work, and later I awakened a few times in the night because my heart was beating too fast. I had a physical a few days after that. My good friend, Dr. Glen Wilson, said I was in excellent physical shape, but he told me to get a good night's sleep, and not work too hard and too long each day. I took that advice.

Then came official, devastating news. At the April 10 stockholders' meeting, Mr. Dansie reported that the United States government was threatening to withdraw the south end of our winter range, so it could be used by the Dugway Proving Ground, a military installation. On April 24, 1950, all livestock owners holding winter grazing permits in that area of south Skull Valley received a letter, which in effect cancelled all grazing on the Dugway Proving Ground. Graziers were invited to attend a meeting scheduled for June 19 and 20, 1950, at Fort Douglas in Salt Lake City, and discuss this situation with F. S. Tandy, Colonel, Corps of Engineers, San Francisco.

Otherwise, we had an excellent spring season. Winter losses were light, and the sheep were fat. Cattle on the range in Skull Valley looked good. I went ahead to Wahsatch to receive the sheep the last five days of April, and Will stayed in Skull Valley to help load and ship the sheep.

On the bright sunny morning of May 13, two young men, Wayne Pehrson and Jerry Johanson from Mt. Pleasant, Utah, were hauled to Home Ranch headquarters so they could drive the saddle horses that hadn't been used on the winter range back to the shearing corral. These were older horses, and all were gentle. Each young man saddled a horse to ride as they drove the other horses. They were to drive the horses south to the corral at Salaratus. Tom and Riley Hale would be camped close by there, with a herd of lambing ewes. The Hales were to take one or two horses. Riley had previously used the horse Jerry was riding, so he asked for that horse, and Jerry gave it to him. Jerry then caught a sorrel gelding, saddled, and mounted. They turned the horses out of the corral, and the two young men headed them south along the road toward the shearing corral.

They had gone about three hundred yards when Wayne had to turn a horse that was not following quite right. When he looked back at Jerry, there was Jerry's horse, but it was riderless. Wayne stopped and looked back a few feet. Jerry was on the ground. He got to him and dismounted. Jerry was badly hurt. Wayne raised him up and called to Tom and Riley. The Hales came on the gallop, but Jerry was nearly gone. He died with his head cradled in Riley's arms. Just then a pickup truck from the Home Ranch was passing, so they flagged down the driver. After discussing the situation, they decided it would be quickest to get the word to Pete Mower at the Home Ranch, and have him notify the proper authorities. This they did. Wayne had a long, tough ride as he drove the horses the remaining eighteen miles to the shearing corral.

I didn't know this accident had happened until later that morning, when I was getting gasoline barrels filled at Bob Ball's service station in the south end of Evanston. Pete Mower, on the way to the shearing corral, stopped when he saw me at the filling station. These things are hard to believe—it takes a moment. I leaned against the pickup truck and reviewed the situation. Jerry's body had been examined and taken to the mortuary in Evanston. Wayne did not see what happened to cause Jerry's death. The horse had been used for several years and was not expected to buck. The herder who had previously used that particular horse had already asked for it when available. Wayne and Jerry had only traveled about three hundred yards with Jerry riding the sorrel gelding. The examination indicated Jerry had died of internal injuries. The horse might have fallen on him, but they did not know for sure.

Pete Mower gave me the telephone number Wayne said to call in Mt. Pleasant, that of a close relative of Jerry's. I called and told this lady what had happened. She could and would go tell Jerry's parents. I told her the name and address of the mortuary in Evanston, and that I would be at this telephone after Jerry's parents had been notified. In less than fifteen minutes, Jerry's father called from Mt. Pleasant. He asked if Jerry got a foot caught in a stirrup and was dragged to death. I said no, and that Wayne Pehrson was with him but didn't see what happened. I explained that they were riding side by side, driving twenty extra saddle horses from the Home Ranch to the shearing corral, when Wayne turned away for a few seconds, and when he turned back, he saw Jerry's riderless horse. I told him the rest of the story of how Jerry died, and said we were terribly sorry about this, as Jerry was a fine young man. Tears were streaming down my face. Speaking with difficulty, I added that I knew how tough this was for Jerry's family. Jerry's father said they would leave for Evanston in a few minutes. I would meet them at the mortuary.

A few days later, at Jerry's funeral service in Mt. Pleasant, I visited with Jerry's folks, and with several people from Mt. Pleasant and Spring City who had worked for the Deseret Live Stock Company in years past. Naturally Wayne Pehrson, Jerry's closest friend, was there. This was a time of extreme sadness for family and friends, and for all who had known Jerry. It would take a long time for those of us at the Deseret Live Stock Company to get over the tragedy of this young man's death out on the range.

A couple of cold snowstorms hit across the grass-covered hills near Wahsatch during lambing, and many newborn lambs were lost. Still, it was a good lambing, and the crew got all but the herd of youngest lambs docked before shearing started. There were twenty herds of ewes and lambs, and five herds of dry ewes, which included the replacement yearlings. When the yearlings were shorn, they did not have to be mouthed, just given a fresh paint brand on the top of the back. Two men did the branding, as the sheep were going through the chute single file. Each brander was instructed to brand every other ewe, that is, brand one, skip one, brand one, skip one. Each brander dipped his brand in the paint pot before branding each ewe. Sometimes the sheep were on a fast trot as they passed by the branders. The men worked fast, but this speeded up the job. I counted as the sheep came out of the chute.

The sheep in this particular herd were being branded with black paint. The men doing the branding were new on this job, or at least to this speeded-up method. I cautioned them to keep the paint pot just above and outside of the chute, so a ewe didn't jump and hit the paint pot, dousing the brander with black paint. Thom wore eyeglasses and was the first brander, stationed about fifteen feet in front of the second man. Both stood outside the chute on the

Felix Albo, left, and Rex Peterson, summer range, 1950.

left side, holding the paint pot in the left hand and the brand in the right. It was a fairly easy job to dip the brand into the paint and apply it to a ewe. Dip, skip a ewe, brand; dip, skip a ewe, brand. The ewes were going at a fast trot, and Thom's paint pot was above the middle of the chute as he leaned over to get closer to his work. A ewe coming down the chute jumped just before she got to Thom, hit the bottom of the paint pot with her head, and Thom got about a gallon of black paint right smack in his face. I stopped the sheep, and Thom put down his brand. He cussed and took off his glasses. What a sight— his face dripping with black paint, and two big white eyes glaring in disbelief. I told Thom to go over to the house and get washed up. What a mess. He was lucky to be wearing glasses, as none of the paint got in his eyes. I asked another man to replace Thom on the branding job, and he took care not to let the paint pot get out over the chute.

Shearing finished on July 3, 1950. All herds were grazing toward their summer ranges. Then, as usual, the crew moved to the summer range. That summer, Bill Cook had his wife and five children living with him on the summer range. They set up camp out south, near the head of Monument Ridge. The Cooks and our family got together several times for dinner. The Cooks had a wonderful family, and they were a pleasure to visit.

At headquarters, all were bedded down one night and sleeping comfortably, when Kathryn and I were awakened by the sound of an approaching strong wind. In a few minutes, we were hit by a real tornado. The cabin where Kathryn and I and the two smallest children were sleeping was torn from its foundation, and one corner kept rising about three feet, then lowering again. Doris was sleeping in the baby crib, which had wheels on the bottom of its legs, so the crib moved a bit as the cabin was raised and lowered. Dale, in a single cot, woke up, but saw that we were right there. Bill and Diane were sleeping on metal cots in a tent to the side of the cabin. In a few minutes they were frantically pounding on the side of the cabin to get in, but the tornado had turned the cabin a quarter of a circle, and the door was not where they thought it was. Fortunately, the cabin stayed on its foundation, but just turned on an angle. A quaking aspen tree, a foot in diameter, had fallen on the tent, but it landed close to one of the metal cots, which held it up at a forty-five-degree angle, and no one was hurt.

Our family spent the rest of the night in the cabin. A hailstorm soon followed the roaring tornado, and the shingle roof of the cabin crackled like a dozen machine guns firing at once. When the hailstorm had passed over, I brought a panful of hail into the cabin. We always remembered that those hailstones were the size of pheasant eggs. The next morning, we looked outside to see dozens of quaking aspens twisted off a few feet above the ground. The new gray Hudson had a large tree across the trunk, which was dented, but otherwise the car was not damaged. In a few minutes, Bill and Diane came running to the cabin to give us more news. "The toilet has blown away," they announced. We laughed and never did find the old toilet. I cut trees all morning to clear the road and parking area. Ed Pederson came with his team, and dragged the trees away. That afternoon, two of the men went with me to a sawmill in the big truck. We returned with a fairly new toilet, a spacious "three-holer."

Earlier, I had proposed to Mr. Dansie and Will that 150 Columbia rams be purchased to replace those culled on account of disease and old age. I explained that Columbias crossed on Rambouillet ewes would produce mostly half-blood ewes, which would work for us. Mr. Dansie saw some good ones at the ram sale in Filer, Idaho, so he arranged to purchase 150, as recommended. In August 1950, a carload of excellent Columbia rams, shipped from Twin Falls, Idaho, arrived at Wahsatch. These rams had the necessary health inspection and certification papers. They were kept in a pasture separate from the main herd of rams for a month. In September, when we gathered the rams for shipping to Skull Valley, the buckherder said he had treated four of the new Columbia rams for an irritation, or infection, of the sheath around the penis. We caught those four rams and inspected the sheath. Each appeared normal. We caught a few more of the new rams, and all were healthy. I cautioned the buckherder to keep on the lookout for anything that resembled pizzle rot, a disease of the sheath. We imported foot rot with the rams from Ohio, and we didn't want anything like pizzle rot coming in on those rams from Idaho.

As lamb-shipping time approached, Will Sorensen supervised the Caterpillar bulldozer program, improving the roads on the summer range so trucks could be driven to the south-end cutting corral, at the head of Horse Ridge. A fleet of twelve trucks hauled lambs from the summer range directly to the stockyards at Ogden. The process worked efficiently at the cutting corrals on the mountain. As the lambs were separated from their mothers, they ran, single file, for a few feet down the chute, and then up the incline into the truck. Lambs were counted as they were loaded, and as soon as a load was completed, an empty truck was driven up.

This was an improvement in lamb marketing. It got the lambs to market within a few hours from the time they were separated from their mothers. These lambs did not shrink in weight, as others did when driven to Wahsatch. Mr. Dansie wrote me a note saying to wait a week after the south-end lambs were shipped before sending more. The lamb market was "thin," and too many arriving within a short time would break the market all across the U.S. We staggered the process of marketing, and things went well.

The fall work went beautifully, followed by five days of shipping to the winter range. Upon arriving in Skull Valley on November 4, 1950, I looked at the herd of rams. It was now evident that there was a serious spread of pizzle-rot disease. Dr. Osguthorpe, the veterinarian, came out. Many of the new Columbia rams had visible symptoms of the disease. Two of the young men living near headquarters and I built two new ram feeding lots. We examined and treated the sheaths on every one of the nine hundred rams, then separated five hundred that were free of the disease into one of the new lots. The remaining four hundred were treated and put into the other new lot. These were treated every three days, until all of them were free of pizzle rot.

Everything else was going just right. In mid-November, as usual, fat cattle were butchered and the beef distributed. December 8–12 was pork-processing time. When it was time to put the rams into the ewe herds on December 15 and 16, we set each ram on his rear end while Dr. Osguthorpe examined the sheath. I told Mr. Dansie that I appreciated the help and expertise of Dr. Osguthorpe. These new Columbia rams, bred to our fine-wool Rambouillet ewes, would sire the kind of ewes we needed on our ranges.

The shipping counts for fall 1950 were as follows:

October 30:	2,581	Voras
	2,610	Baca
	<u>2,688</u>	Jaramillo
	7,879	
October 31:	2,738	Rex
	2,503	George
	2,512	Dell
	<u>2,618</u>	Romero
	10,371	
November 1:	2,630	Max
	2,680	Manual
	<u>2,682</u>	Dan
	7,992	
November 2:	2,545	Tom
	2,388	Larry
	<u>2,504</u>	Fred
	7,437	
November 3:	2,504	Joe
	<u>2,304</u>	Ben
	4,808	
	38,487	to Skull Valley
Also shipped		
November 3:	<u>2,025</u>	ewe lambs to California
	40,512	
Rams	<u>900</u>	
	41,412	Total to winter

18
Moving Up

The inventory count of ewes for December 31, 1950, showed 38,222 in Utah and 2,025 in California, for a total of 40,247. The cattle inventory was 5,058. The weather was favorable, an easy winter. At the annual stockholders' meeting in Salt Lake City on March 26, 1951, Mr. Dansie reported that the sheep were in the best condition he had ever seen. At this point, Will and Vera Sorensen were in Florida for several months. He was helping with the land and water development on the Florida property. I lived in a new rubber-tired campwagon at the ranch, but drove home to Grantsville three or four nights each week. I needed to keep close to Kathryn and to each of the four children.

At the end of March, when I came home one evening, Kathryn told me that we were going to have to move out of our house when school was out, as Denny had a chance to sell it. I replied that we did need a little more room, so I'd discuss it with Mr. Dansie. A few months ago, he had said he would like us to move to Salt Lake City when the time was right, and this perhaps was it. A couple of days later, Mr. Dansie suggested that we start looking for a new home in one of the areas south of Salt Lake City. That location would be most convenient for our future with the Deseret Live Stock Company. So while I was busy with the sheep operation during the spring, Kathryn loaded all four kids into the gray Hudson and spent part of several Saturdays becoming familiar with the homes, schools, and neighborhoods adjoining Salt Lake City on the south.

One day Bill Watts, Art Krantz, and I, along with the herders and camptenders, were on the west side of Cedar Mountain, and had just finished putting two herds through the corral. I operated the dodge gate, cut out the thin sheep, and at the same time counted the herd as they came single file through

155

the chute. Bill and Art were each driving one of our large farm trucks, which were equipped with a double-deck for hauling sheep. I had the Jeep pickup, which had a stock rack. We loaded all three trucks as fully as possible, and still had five ewes too many. We put two ewes in the front seat of the truck with Bill, put two in the front of the other truck with Art, put the last ewe in the front of the pickup with me, and headed for U.S. Highway 40, six miles to the north.

We had to travel east on this major highway for twenty-five miles before turning off to go south up Skull Valley to the ranch. Art was in the lead, followed by Bill. I brought up the rear. Traveling along the highway, people in approaching automobiles took a surprised look at Art in the lead truck when they saw the ewes in the front seat with him. Most people had a big grin on their faces as they passed Bill and the ewes in the second truck. I reached in back of the seat, grabbed Dale's little old straw hat with a string going under the chin, slipped it onto the ewe in the front with me, and pulled her over a little closer. This caused chuckles and hand waves from passing motorists. When we reached the ranch and were unloading, we laughed about the surprised looks of the people we had passed. Bill said that the one old ewe closest to him was chewing her cud and looking at the approaching traffic. He put his arm around her and waved at the people. Art was embarrassed and hoped no one passed that knew him.

Spring came early on the desert, feed was plentiful, and sheep death losses were not too high. Having plenty of water for the sheep after the snow melted was always a problem, particularly on the west side of Cedar Mountain. Although four herds could drink at the well on the west side, the new tank and troughs at Quincy Spring were a big improvement, as it kept one herd away from the well for three weeks. I measured the flow of water at Cedar Spring, which was on the west side, but on its south end—about twenty-eight miles south of Highway 40, midway between Rattlesnake Point and Wig Mountain. The spring was sufficient to water one large herd of 2,800 ewes coming to water every other day. I proposed to Mr. Dansie that I would like to install a 3,000-gallon storage tank and a set of watering troughs at Cedar Spring. He told me to go ahead if I was convinced it would water one herd. In April, three of us installed the tank and 150 linear feet of metal troughs. This was another big improvement for watering a herd. It spread out the herds and helped protect the range.

We shipped off the desert, from Timpie, the first five days in May. I went ahead to Wahsatch to receive the sheep and get all the herds headed to their designated lambing ranges. By now, Kathryn had located what she figured was the best new home we could afford. It was in the Granger area, an attractive

At the Deseret Live Stock Company's summer range, 1951, front, from left,
Kathryn Frischknecht, son Dale, and Marj Fidler; in the window, from left,
Carol and Virginia Fidler, Doris Frischknecht.

suburb south and west of Salt Lake City. I looked it over with her, and we
closed the deal. Granger was a rapidly growing community surrounded by
rich farm land. Broad fields of wheat, barley, oats, and alfalfa stretched in
every direction.

This was our first purchased home. It was in an attractive setting, on a
rounded corner lot. It was of dark red brick, with a heavy tile roof that would
last a lifetime, and a natural gas furnace. The L-shaped living and dining room
had a fireplace, which we knew all the family would enjoy. A kitchen, bathroom, and two bedrooms were also on the main floor. The home had a full
basement, with windows above ground level. The house had just been built,
and inside painting had not been done, so Kathryn was able to select the col-
ors she wanted in each room. When school was out for the year, our family
vacated the home in Grantsville, said some tearful goodbyes, and told friends
we would keep in touch. We moved the furniture into the new home, and then
Kathryn and the kids joined me at spring sheep headquarters, the shearing
corral near Wahsatch.

Tom Judd had been cooking for the headquarters crew at the shearing
corral since we moved there in late April. He knew how and took a lot of
pride in his cooking. Shearing started June 18, 1951. The same Texas crew

had contracted to do the job, but they arrived a week later than anticipated. Because of the late start, shearing continued right through the Fourth of July, but no one complained. With the work finished at the shearing corral, it was "on to the summer range." We went to the summer range with twenty-one large herds of ewes and lambs, and four large herds of dries, including the yearling replacement ewes.

Soon Kathryn and I and our company of four children drove to our new home in Granger, so we could do the laundry and take comfortable baths. It was timed so we could be home on a Sunday. We attended church at the old yellow brick building at 35th South and 32nd West in Salt Lake City. New homes were springing up. At summer headquarters, all had enjoyed those beautiful sunny days of July and August. Now it was time for Kathryn and the children to move home for the beginning of school. Young Bill was starting fourth grade, and Diane was starting third. Dale and Doris were too young for school. When it was possible for me to spend a day or two in Salt Lake City, we worked on landscaping our home. We leveled the yard, hauled in topsoil, planted lawns, fertilized, and sprinkled. We planted several trees, shrubs, rose bushes, and other flowers. We purchased groceries at ZCMI and laid in a large supply of cases of canned vegetables, jam, honey, fruit, and whatever other food came in cases. I hauled a load of fireplace wood in the pickup when I came back from trips out on the range. Kathryn and I were happy to be making an investment in a home and were thankful for four healthy children.

Tom Judd had become acquainted with a lovely, widowed, middle-aged lady, Ethel "Jim" Schultz, when both were visiting Tom's aunt, Goldie Johnson, in Grantsville in 1950 and 1951. They enjoyed each other's company, and although they discussed marriage, she had a contract teaching school in Hood River, Oregon. In August 1951, she wrote to Tom from Oregon, telling him that she had decided to accept his proposal of marriage. She resigned her position in Hood River, and said she would meet him in Salt Lake City on September 1. Tom was elated. He had been a lifelong bachelor, and now he was going to get married. He planned to be gone for two weeks, and then would return to Wahsatch to help with the fall work. Tom and Jim were married in Elko, Nevada, on September 4, 1951. After a few days' honeymoon, they rented living quarters from Aunt Goldie. Tom returned to his job with the Deseret Live Stock Company, and Jim went to work at the Tooele Ordnance Depot, about twelve miles from Grantsville. Tom was a new man when he returned to his job, happy, enthusiastic, and in love. Each person on the outfit warmly congratulated Tom.

The fence crew had been discontinued in late August, so the boys could return to high school. Lamb shipping started, as usual, on the south end in

Tom Judd and Ethel "Jim" Schultz were married in
Elko, Nevada, on September 4, 1951.

mid-September, and now all the lambs were trucked from the summer range
directly west to the Ogden stockyards. The crew of eleven trucks made two
trips each day, and the lambs were weighed in Ogden a few hours after being
separated from their mothers. This was a great improvement.

I closed up the summer headquarters in early October and moved my camp
to the dipping corral at Salaratus. Here, the crew put all the herds through the
corral, made up the winter herds, and rebranded them with fresh paint. We
did not dip the sheep in water, but planned to apply dry insecticide dust at the
shearing corral, during eye-shearing time in late October.

Mr. Dansie came up for a conference with Bill Cook and me. He told us
that he liked the way our 2,000 ewe lambs grew last winter in California, but
said that 2,000 lambs per herd were too many for this winter. He instructed us
that when the sheep were at the shearing corral, we were to separate out three
bunches of 1,900 each, which one man could take to the Imperial Valley.

Tom Judd handled the cooking at the shearing corral. The well-organized
crew vaccinated all the replacement-ewe lambs to prevent soremouth, and
accomplished the eye shearing and insecticide dusting before we segregated
the ewe lambs. The days were cold and clear, and the prevailing autumn breeze
across Wahsatch's treeless, grass-covered hills was often a chilling wind. One

cold afternoon, Bill Watts and I were on one side of the chute vaccinating ewe lambs, and Bill Cook and Bill Gustin worked from the opposite side. After a couple of dusty hours, things had quieted down to just everybody doing his job. No one was talking at the moment when Bill Watts straightened up and took a long look in an easterly direction. "Look," he said, pointing to the east, "Here comes a man from Jerusalem." All of us stopped working and straightened up to look. Silhouetted broadside against the sky was one of the herders, riding a mule down the top of a ridge. The mule traveled with its head straight out in front, on a level with its back. It looked just like a scene from Biblical times in the Holy Land. Bill Cook said he was glad Bill Watts was familiar with the Bible.

As we finished each chute full of sheep, one man turned the switch to start the dusting machine, situated at the far end of the chute. As the sheep were released from the chute, they passed through rotenone dust, which was blown, under high pressure, into their fleeces. This method of dry dusting was easier than putting sheep through the dipping vat, and it was very effective in eliminating external parasites. One problem, though, was that the dust also penetrated the clothes and hair of the men, and we carried a strong odor.

One evening after supper, two of the men rode with me to Evanston to take a shower at the YMCA. We stopped for a coke at Freeman's before going to the Y. I noticed that the waiter looked at us and sniffed the rotenone odor, so I said that we were going over to the YMCA and take a shower. "Good, go now," he said, and then smiled and went by without serving us. We went over and showered, taking care to give our hair a thorough washing. That rotenone odor was not easily overcome. After the shower, we went back to Freeman's. When the same waiter came to take our order, he sniffed and said that he thought we were going to go over and take a shower. I replied that we did, but he said that we still stank, just like when we came in a little while ago. The waiter wasn't smiling, and neither were the other customers. It looked like an opportune time to say goodnight and go back to Wahsatch, without a coke. When we got up to leave, the waiter said, "Be sure and come back sometime." He then said he wished we really would take a shower.

In a few days, we had loaded 5,700 replacement-ewe lambs onto the Union Pacific, destination Imperial Valley. I was particularly pleased, as we would have only about 3,000 of the biggest ewe lambs wintering in Skull Valley. Regular shipping to Skull Valley was on November 7, 8, and 9. Bill Cook wanted to hold the other three south-end herds in the Heiner's Canyon area for another two or three weeks, that is, until the snow got too deep for them to graze any more. In late November, Mr. Dansie called me and said that the three herds still at Heiner's Canyon were in deep snow, and that I needed to take a couple of our men and go help get the sheep loaded for shipping to Skull Valley.

Will Sorensen came from his home in Spring City to lend us a hand, and Bill Watts, Tom Judd, and I met him in Salt Lake City. We stayed at a motel in Coalville en route to the loading corrals at Emory. It would be easier to ship from Emory, which was twenty miles west of Wahsatch, at a lower elevation and thus with much less snow. The next day we met Bill Cook and loaded the seven wagons onto the two gondola cars before noon. We then loaded the three herds. The final jobs were to load the fourteen horses, then sign the railroad's bills of lading. The train was ready to head for Skull Valley by 3:30 p.m. Things went well.

The men enjoyed working with Will Sorensen again. He was breaking away from the Deseret Live Stock Company a little at a time, but he was still active on the board of directors. The operation of the company was the business concern foremost in his mind. He enjoyed retirement with Vera at their home in Spring City. She was a pleasant, happy companion and took good care of him. This helped him to maintain his quiet dignity, and he was most considerate of her. Both had a keen sense of humor and enjoyed telling decent stories of the old days. Will, Vera, Kathryn, and I were looking forward to a great occasion in about a week.

THE BEST OF TIMES

DURING EARLY NOVEMBER 1951, BEFORE shipping to the winter range, Mr. Dansie had confided to me that in a month the company would like to send Kathryn and me to the National Woolgrowers' Convention, scheduled for December 4 through 7, in Portland, Oregon. They would bring Will and Vera Sorensen out of retirement in Spring City to accompany us, and we'd travel on the railroad. Kathryn and I spent an afternoon in Salt Lake City finding respectable attire.

We invited Will and Vera for dinner at our Granger home the day of departure, and it was a pleasant reunion. All four children were happy to see Mr. and Mrs. Sorensen again, and each child gave them a loving hug. Will gave each one a shiny silver dollar. He had done this several times before. Young Bill told Will that he had saved every silver dollar Will had given to him. This pleased Will and Vera. My mother came from Manti to take care of our kids while we were gone. She had an easy way with children, having raised six of her own, so all of us felt good about her being there.

At the railroad station in Salt Lake City, we were pleasantly surprised to see several dozen Utah people heading for the convention. We rode to Ogden, and there joined a "convention-special train" of sheep producers coming from across the United States. This was a good time for relaxing and getting acquainted. The club cars were overflowing. The convention was held in the Multnomah Hotel, but due to the large national attendance, it was necessary to have people stay in several hotels in Portland. The Sorensens, Kathryn, and I were housed at the New Heathman, a few blocks south of the Multnomah.

I again served on the Lamb Marketing Committee, which was chaired by Ervie Williams, manager of the North Portland Union Stockyards. When he entered the room just before the meeting started, we recognized each other from the 1941 college livestock judging contest, held at the Pacific International Livestock Exposition in Portland, when I was a member of Utah State University's livestock-judging team. He was one of the executives in charge of the show. We shook hands and had a brief visit. My new, all-wool, charcoal grey, double-breasted business suit from Arthur Frank's in Salt Lake City was a distinct asset.

Just before the next session of the convention was to start, a tap on the shoulder caused me to immediately turn around. To my pleasant surprise, it was Dr. Fred F. McKenzie, who had been my major professor at Utah State for both my B.S. and M.S. degrees. He was now the head of the Animal Science Department at Oregon State University in Corvallis. We had a pleasant reunion, and Dr. McKenzie introduced me to Dr. Ralph Bogart, a geneticist teaching and doing research at Oregon State. I had learned a great deal from Dr. McKenzie, and it was good to see him again. It had been eight years since my M.S. degree was finished. Dr. McKenzie had been a major influence in getting me to work on a project at the Deseret Live Stock Company during the 1943 shearing season. That resulted in my eventual employment at the company in 1946. The meeting was about to start, but Dr. McKenzie told me that it looked as if things were working out well for me at Deseret. I told him that Walter Dansie was a true gentleman and an exceptionally able general manager. Will Sorensen had taught me how to run the forty thousand sheep. It was a great experience, and I thanked Dr. McKenzie for helping me get acquainted there.

That evening, Kathryn and I went to dinner with Will and Vera Sorensen and two other couples from Utah: Mr. and Mrs. Parley Madsen and, from our hometown of Manti, Mr. and Mrs. Wilford Wintch. Charlie Redd, from La Sal in southeastern Utah, also joined us. I had graded 25,000 fleeces at the Redd ranch in 1943 before going to the Deseret Live Stock Company, so we were well acquainted. We Utah drylanders enjoyed a seafood dinner at a high-class restaurant, and all of us had a good time visiting. Will knew Wilford and Parley from when they were all young men. Will and Wilford had been almost like brothers, working together for Wilford's father, Jacob Wintch, in Manti. When the waiter brought the check, Will reached for it and said, "I'll handle this." This was a little surprising. However, it just showed the generous side of Will. He was always a quiet and reserved gentleman and had the respect of everyone.

Will told them that when I came on the job, it relieved him of the anxiety of running the sheep, and his life became more enjoyable. Will was now fully retired, and he and Vera were happy to live in their Spring City home. However, Will remained a major influence on me as he continued to give me the benefit of his long years of wisdom and experience in dealing with nature in a hard environment.

During the convention, Kathryn and I became acquainted with a young couple from North Dakota. That terrible January and February of 1949, the winter of the deep snow, was still on everyone's mind. The young man from North Dakota told us that they were snowed in, and that their ranch was several miles from the nearest neighbors. Another young couple who lived five miles distant were also snowed in, and they were expecting a new baby. His wife was concerned, he continued, about the young neighbors who were almost ready to have the baby. She asked him to bundle up, put on snowshoes, and go over to see how the neighbors were faring. He put on all-wool clothes, then sun goggles, and stepped into his snowshoes. His wife helped him put on his big buffalo-robe overcoat, tanned with the hair on. Tears came to his eyes, and he paused for a moment. He then said that he started out in a direct line for the neighbor's house, because the snow covered all the pasture fences. It was a clear, cold day, and he was the only critter doing any traveling. When he got halfway to the neighbors' place, a large transport airplane flew over at low altitude, probably a thousand feet. It banked, then circled back in his direction. As it went by, a bale of hay came down from its ponderous belly— another buffalo saved.

When we returned from the convention, Will and I had a conference with Mr. Dansie in the company's office in Salt Lake City. Everything was under control. Will and Vera returned to Spring City, Kathryn and I to our home. I left home at five the next morning, in order to be at the ranch in Skull Valley for breakfast at six thirty. Bill Cook, Glen Hess, and I made arrangements to butcher twenty-one hogs the next day. Mr. Dansie came to Skull Valley for a conference with us, and to look over the rams before they were to go out to the ewe herds. He told us that he had purchased a set of ninety young rams, which would be delivered to the Home Ranch so they could breed that herd. That should make over nine hundred rams in our year-end inventory. He was pleased with the condition of the rams and was glad that we weren't fighting foot rot or pizzle rot this year. He had a clean bill of health on the new rams, and the veterinarian said he was sure of it.

Mr. Dansie was back in a few days and brought the Christmas checks, one-twelfth of what each employee had earned during the year, an extra month's pay for full-time employees. Again I told him that each of us—every herder, camptender, and cowboy—appreciated this gesture of thanks from the company. It generated loyalty and goodwill on the outfit.

Mr. Dansie, Bill Cook, and I went over the total number of replace-ment-ewe lambs that would be yearlings in the spring. Mr. Dansie said that we had 5,680 in California and, allowing for some loss, about 3,640 in Skull Valley. That would give us 9,000 yearlings, a great addition to our herds. The year-end inventory for December 31, 1951, and January 1, 1952, showed the following:

Ewes on winter range	Skull Valley	34,073
Ewes on winter range	Home Ranch	2,755
Yearlings in California		<u>5,680</u>
Total Ewes		42,508
Rams		<u>942</u>
Total Sheep		43,450

The winter was going smoothly. Rams and thin ewes were taken out of the herds at the end of January. In late February, Mr. Dansie called me and said that the main buildings at the new dairy over on the Crane Ranch, west of Evanston, burned to the ground on Sunday, February 24. The cows had to be taken to the Home Ranch barns until the others could be rebuilt. He reported on a few details, then added that Will Sorensen had helped oversee the construction when the dairy facilities were built a couple of years earlier. When the weather warmed up, the company would ask him to come back and do it all over again. Mr. Dansie said that the loss of the dairy was not as devastating as what the government was taking away on our winter range. When the U.S. government established Dugway Proving Ground—in the southern part of Skull Valley and on the south end of Cedar Mountain— the Deseret Live Stock Company lost winter range for 9,250 sheep when this land was withdrawn from grazing. He continued that between then and November, we were going to have to find new winter range for four winter herds. Corrals at strategic locations on the winter range had made it easy to put herds through the corral each month and to truck a few thin sheep to the ranch for extra feeding.

The annual stockholders' meeting was held on March 24, 1952, at the Hotel Utah in Salt Lake City. Deseret Live Stock Company officers and directors at that time were: Henry D. Moyle, president; James D. Moyle, vice-president; Walter Dansie, secretary-treasurer and general manager; and James H. Moss, Edward O. Muir, William H. Sorensen, and R. John Moss, directors. At that meeting, it was reported that the company's net income for 1951 was $636,864, and that the federal income tax paid that year was $336,000. All stockholders received a 6 percent dividend in 1951. I told Mr. Dansie that over $300,000 was a lot of money for one ranch to have to pay in income taxes. He replied that that was because the company was on the accrual system, which President Moyle, in the past, figured was the best system to use.

Larry Vialpando and his team Tip and Shorty at the south well in Skull
Valley getting a barrel of water, April 1952.

The company had long owned the right to process twenty second-feet of
Great Salt Lake water to harvest salt. This right had never been used until the
previous year. In an attempt to diversify and use all assets, the Deseret Live
Stock Company built a salt plant on the south shore of Great Salt Lake, near
the small community of Lake Point. The report for 1951 showed 25,000 tons
of salt had been harvested and processed that year.

Before adjourning the meeting, President Moyle took this occasion to "pay
high tribute to the work being done by Mr. Dansie and by Dean Frischknecht,
in charge of the sheep operations under Mr. Dansie's direction."

Early in the morning one spring day, I drove up to Bill Watts's camp,
located at the well on the west side of Cedar Mountain. I was walking toward
the camp when an attractive young lady looked out from the camp and softly
said "Good morning." Then she told me that the baby was sleeping. She
could see I was shocked, so she quickly added that her husband and Bill were
in Bill's truck and had just gone to look at a herd of sheep. They would be
back in a few minutes. She added, "You must be Dean," and said that Bill
told them I was coming that day. I asked her when they had arrived, and she
told me Bill brought her and her husband and baby to his camp the previous
night. Her husband was just discharged from the Navy, in California. They
had bought a used car and figured on driving it to their home on the East

Coast. They were having car trouble, but last night it finally limped in to the Riddle Brothers' service station at Low, over on the highway. Bill was there, and when he saw their predicament, he invited them to use his camp for the night. I told her that I was glad he did. They locked their car, she continued, and Bill brought them to his camp. He stayed in the cabin by the well. She was going to fix breakfast for the men when they got back. Bill had told them he would take them back to their car, and they would decide what was the best way to go on home.

She and I chatted for a few minutes, and then the two men came back. Bill introduced me to her husband, a pleasant and nice-looking young man. They were a fine young couple, with a small child, and they could use a little help. The young man climbed into the wagon to give his wife a hand. As Bill and I walked a few feet to the pump, Bill told me that he had figured he should get these young people and their baby into the warmth and shelter of his camp for the night. That way they could get a good night's sleep in the bed, and that was a lot better than spending the night in their car. They were good people, and Bill did the right thing.

After breakfast, Bill and I finished our sheep business, and then he drove the young people to their car at the Low station. I followed and stopped at Low to see what should be done. The young couple had decided it would be best for them and the baby to go by bus to the East. It would be safer and surer than trying to fix their car. Bill said that he would box up whatever they couldn't take on the bus and send it to them. He purchased their old automobile for cash and made sure they had enough money to get home.

The young man and his wife were happy and much relieved. They wanted to know how to contact Bill, so he said, "Just write to me at Deseret Live Stock Company, Grantsville, Utah." They listened closely, and asked again for the address. Again he said, "Deseret Live Stock Company, Grantsville, Utah." She said she had his name and the address written down, but asked him to please repeat again, slowly, the name of the company. So he drawled out "Des-er-et Live-Stock Company." She had it all written down by now. A few days later, Bill got a thank-you letter from them, addressed to "Bill Watts, Desert Rat Livestock Company, Grantsville, Utah."

The Dugway Withdrawal

I LEFT A DAY EARLY to receive the sheep at Wahsatch, and stopped in Salt Lake City to give the sheep counts and the schedule for shipping at Timpie, to Mr. Dansie. He told me that when he saw some of the herds a couple of weeks ago, they looked good; the winter loss was much lighter than in the old days. I replied that we now had four corrals on the west side of Cedar Mountain and five corrals in Skull Valley. We could get a big truck to all of them, except the corral at Cane Spring, on the west side of the south end. With our trucks, it was easy to take out thin sheep and get them on extra feed. However, we still lost a lot of sheep to coyotes.

We talked over a few more things, and I told him I needed to pick up some new pocket knives. In previous years, two or three, or maybe up to six, ewes died each day while being shipped. We dragged the dead ones out of the railroad car, so they could be skinned. The pelts had long wool, and I made sure we got them skinned, but when I asked some of the camptenders to help skin the ewes, they said they didn't have pocket knives. Mr. Dansie instructed me to go to our wholesale house and get whatever we needed. The Tree brand knives were some of the best, and they came six to a carton. I decided to get three cartons and then sell a new knife to whoever wanted one, with the understanding we'd withhold the $3.75 from their next paycheck. At the unloading corrals at Wahsatch, new knives were picked up each day. When Bill Cook arrived, he asked for my last carton of six and said he would keep track. I got on the phone and ordered two more cartons.

Snow drifts on the road from Wahsatch to the shearing corral made traveling difficult, so Riley Hale and I camped at the unloading corral at Wahsatch. Riley, in his sixties, was doing the cooking one day, while preparing dinner, he complained that he was losing strength in one hand. Fearing he might be in

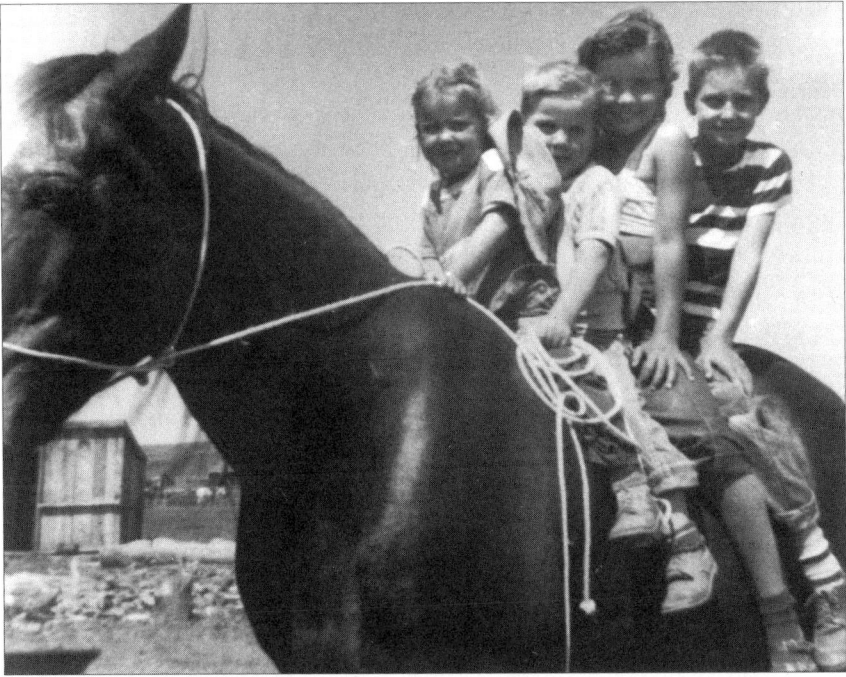

Doris, Dale, Diane, and "Bill" Dean Frischknecht on Old Bally, May 1953, at the Deseret Live Stock Company's shearing corral.

for trouble, he asked to be taken to his home in Grantsville. We kept in close touch, but Riley went downhill fast. I attended his funeral a couple of weeks later. Riley was one of the "old breed," a true gentleman of quiet decency and dignity, considerate of others. He and his wife had raised a fine family. He knew how to work, could see what needed to be done, and went ahead and did it. He had the love and respect of everyone. We lost a good man, and a good friend.

The California yearlings returned to Wahsatch in mid-May. Death loss was low, because they had been pastured in alfalfa fields away from most predators. As Bill Cook and I weighed each draft of yearlings across the scale, we determined their average weight to be nearly one hundred pounds, which was excellent. At least two hundred, the Delaine Merino crossbreds, were too small. However, the positive results showed our program of wintering ewe lambs in California was a successful endeavor. Mr. Dansie deserved a lot of credit for this innovative operation.

During May, snow fell intermittently every few days, although the sun melted most of it during the long afternoons, and the four children—Bill, Diane, Dale, and Doris—came to our house at the shearing corral as soon as school ended. Those were great days, and it was good to be back out on the range. As usual, docking the lambs started the last week in May. Because of the cold storms during lambing, a lot of newborn lambs had died, and now the docking counts of lambs were less than Bill Cook and I had hoped. We reported to Mr. Dansie that we had docked 24,460 lambs. In most herds, we had over a hundred young ewes whose lambs had died. Mr. Dansie said it was still more lambs than we used to have.

A letter to the Deseret Live Stock Company from the CPA firm of Wells, Baxter and Miller, dated June 3, 1952, gave the following expense account regarding the cost of wintering the 5,680 ewe lambs sent to California in November 1951.

Freight—Utah to Imperial Valley		$4,842.76
Feed and Care:	Nov. 8–10 to Nov. 30	7,165.80
	Dec. 1–31	9,735.55
	Jan. 1–31	9,696.49
	Feb. 1–29	9,035.53
	Mar. 1–31	9,607.52
	Apr. 1-May 8	10,843.00
Freight—California to Utah		6,035.37
		$66,962.02

Mr. Dansie reported that those replacement-ewe lambs grew well during the winter, and the death loss was low. It was a break-even proposition. The total cost, including freight, was $11.78 per head for those five months. We were also sparing the winter range. Wintering these ewe lambs in California had been the best alternative.

Shearing got underway June 17, 1952. The same crew from Texas did the job. They wanted to shear every day, including Sundays. A wool buyer, representing an eastern company that had previously purchased the wool, came to the shearing corral. That company wanted the wool again, and their buyer called me aside for a conference. He said that he understood I was an experienced wool man, with a master's degree in animal science. He couldn't figure out why we would breed fine-wool Merino rams to coarse-wool, quarter-blood Romney ewes, and why we would breed those coarse-wool quarter-blood Romney rams to our fine-wool ewes. He showed me a photo of a cross-section of wool fibers from one of our crossbred ewes. There were quarter-blood fibers in among the fine fibers, and that was going to lower the value of our wool. This sample couldn't be graded as fine, and it wasn't quarter-blood. It was a mix of grades, side by side, throughout the fleece. I told him that I knew this would happen. I'd persuaded Mr. Dansie that we should use nothing

Kathryn Frischknecht, with .22-caliber rifle, target
shooting, summer range, 1952.

coarser than Columbia rams, and nothing finer than Rambouillets. It would
take time to straighten out this situation, but we were now on the right track
for raising our replacement ewes. I told the buyer my predecessor was the one
who wanted to cross Merinos with Romneys, and I knew it was too severe. I
told him to keep coming for the wool, as it would improve each year.

Visitors came most afternoons. They just wanted to talk a little, see the
shearing, and marvel at the huge herds of ewes and lambs as they came into
the corral or moved out. The three men loading heavy bags on the wool truck
drew a lot of interest. Kathryn usually wore blue jeans in the morning, but
changed into an attractive housedress for the afternoon. She and Diane and
Doris then walked over to the shearing corral to chat with the visitors. She
told me that many of those people had some tie to the company, either as
stockholders, previous employees, or a relative of some famous person who
once worked for it. They told Kathryn how privileged we were to be here. She
thought they'd forgotten, or didn't know, about carrying water from the well,
and using the outside toilet.

It took eighteen days of shearing to complete the job. They finished up
late in the afternoon on the Fourth of July. The crew had worked faithfully
and hard. However, on account of rain, there were three short days. As usual,
Mr. Dansie wrote checks to settle the shearing account and paid off the temporary employees.

The summer of 1952, looking from the south, when a new frame house was being built where the old log cabin, "the big silver house," had stood.

During shearing, Mr. Dansie asked me to go to Salt Lake City on June 20 and swear before a notary public with regard to the loss of winter range. The government had agreed to recompense the ranchers for whatever was determined by federal appraisers to be the damage sustained by each grazier. Those federal appraisers had a difficult job. Appraised damages varied, the highest being $10 per head to some ranchers. The Deseret Live Stock Company was offered $35,000 as their appraised loss. Mr. Dansie, the directors, Bill Cook, and I all said this appraisal of less than $4 per head was too low. We sought some means of getting this figure raised to something more in line with the actual damage sustained. My statement noted that the Dugway withdrawal of grazing land caused the Deseret Live Stock Company to lose winter range for 9,250 sheep. The company also lost the use of five privately-owned watering places, two of which were leased from neighbors. A similar sworn statement was made by Will Sorensen. He said he had been sheep foreman for many years, up until 1946, at which time he became semi-retired but was always in close touch.

Once shearing was done, Bill Watts, Tom and Jim Judd, and the fence crew moved to a convenient location close to their work. Tom would cook for the fence crew. Our family moved to summer headquarters, and Bill Cook camped at the head of Monument Ridge, on the south end. Will and Vera Sorensen moved over to the burned-down dairy. Later, Will would help Mr. Dansie look for a new winter range. On the summer range, I planned to finish the new house, but a visit by Mr. Dansie changed that plan.

21

MAJOR CHANGES

AT SUMMER HEADQUARTERS, MR. DANSIE informed me that the Deseret Live Stock Company was in the beginning stages of changing ownership. He had been told we should not put any more money into this new house. The buyers were successful businessmen, mostly from Salt Lake City. They were organizing a syndicate of a dozen or more investors to purchase the company. Will had previously told me that occasionally a few shares of company stock changed hands, at $14 per share. Mr. Dansie said that the new buyers offered President Henry D. Moyle nearly twice the going price. They would pay $28, give or take a few cents, for the stock owned by the Moyle family. Henry's brother, Walter, was a lawyer in Washington, D.C., and he wanted to sell. He alone owned one-seventh of the company. Henry and the other Moyles owned about one-seventh; together they all owned close to one-third. Henry had told the buyers that the Moyles would sell, providing all other stockholders were offered the same price offered to him.

We visited about many things, particularly the future of the company. Mr. Dansie proposed that I take a week's vacation. A couple of days later, Kathryn and I loaded the four children into our new green Hudson Hornet, purchased in Manti from Art Nell, for a trip though Yellowstone Park, and then on to Billings, Montana, to visit Bob and Genevieve Holloran and their son Bobby. We stopped to read all the historical markers, and relived some of the Old West action of one hundred years ago. Back on the summer range, our family had one of the greatest summers ever. We rode horses several times. All the family made trips with me in the truck as we picked up butter and flour at the Home Ranch, and visited the different herds, herders, and camptenders. We made several trips to Wahsatch to mail letters, pick up mail, and haul bags of sheep salt to the mountain. I helped Kathryn and the children move home for the start of school.

Genevieve and Bob Holloran and Kathryn Frischknecht. The Hollorans and
their son Bobby stayed overnight with us on the summer range. They are
Marine Corps friends and lived in Montana.

After considerable time was spent looking for a range that would accommodate four herds—a minimum of ten thousand sheep—it was decided to
purchase the Pilot Peak Ranch in eastern Nevada. This once was a part of
the famous old "UC" (Utah Construction) ranch. The east boundary was
the Utah/Nevada state line. The ranch line was at the top of the pass a mile
west of the city center of Wendover, a bustling, isolated desert metropolis.
The ranch was a country of far horizons and wide open spaces, with historic,
snow-capped Pilot Peak towering 10,720 feet near the northeast boundary,
twenty-five miles north of Wendover. The peak got its name for being a pilot,
or landmark guide, for early-day mountain men and pioneers. The Donner-
Reed party in 1846, and later travelers, used the peak as a long-distance friend,
becoming more visible each day as they plodded forward. About half of this
ranch had been a railroad grant. Ownership of the land was like a checkerboard, that is, every other section of 640 acres was private, and the next section federally owned, under jurisdiction of the BLM. Actually, 118,840 acres
were included in this semi-private holding, and several thousand acres of federal land adjoining on the south were used on a grazing permit. BLM officials
rated the ranch grazing capacity as able to carry 11,200 sheep for six months,
from November 1 though April 30.

The purchase price of the ranch was $215,000. The land lay on both sides of east-west U.S. Highway 40. Ranch headquarters consisted of a modest white frame house, corrals, and outbuildings, on one hundred acres of irrigated land west of the landmark peak. Snow from the peak supplied both irrigation water and underground domestic water. This new setup was going to complicate the Deseret Live Stock Company's winter operation. It was more than a hundred miles from the ranch headquarters in Skull Valley to the ranch headquarters in Nevada. It was within operating distance for me.

At a special stockholders' meeting on September 18, 1952, at the Hotel Utah in Salt Lake City, it was announced that the appraised value of the Deseret Live Stock Company was $5,936,429, a value of $73.49 per share for 75,000 shares. In addition, the company had invested $750,000 in the salt plant at Lake Point. It was also reported that the prospective new owners had already acquired 40 percent of the company stock and wished to have three of their own people on the board of directors. This request was granted.

Because the winter sheep operation in Skull Valley was drastically curtailed on the south end by the loss of four herds, Bill Cook terminated his employment at the Deseret Live Stock Company that fall. We had worked well together. The three remaining winter herds on the south end were again my responsibility. Bill Watts camped at his usual place, on the west side of Cedar Mountain. We could handle the ten herds in Utah. I planned to go to Nevada every few days. A sub-foreman was to be hired to live at the Pilot Peak ranch in Nevada, and he would be visited every few days. Mr. Dansie hired Wallace Denison. He was an experienced man, thirty-five years old; his wife was a little younger. They had no children.

Will Sorensen came, particularly to help get the four herds properly located in Nevada. He, Mr. Dansie, and I agreed that the four Nevada herds should be shipped two herds per day, for two days, during the first week of November, before shipping the herds going to Skull Valley. The men heading for Nevada were told about the new range. All looked forward to the experience.

At Wahsatch, Mr. Dansie counted each carload of sheep onto the U.P. Railroad. Will Sorensen accompanied me to Wendover to receive the four herds. The herders, brands, and counts were:

Lujan	Red	3	2,905
Lucero	Red	•	2,848
Herrera	Red	1	2,836
Maestas	Black	0	2,835
Total to Nevada			11,424

Author's Collection
A matched team on the winter range, ready to move out, November 1953.

Will advised me to unload the wagons, horses, and camptenders at the U.P. stockyards in Wendover. The herders and dogs could ride with us in my pickup to where the sheep got unloaded out on the range. He said Walt Dansie and he went over all this operation with Wallace Denison.

Will and I met Wallace at the stockyard. Wally, as he was called, and I had known each other when growing up near Manti. It was a pleasant reunion. Wally reported that there was plenty of snow for the sheep on the north slopes. The special sheep train pulled off the main track onto a railroad siding out on the range, ten miles west of Wendover. There were no corrals. We simply opened the railroad car doors and put up a wooden-plank unloading platform for the sheep to walk down onto their range. This was a great saving on sheep, men, and dogs. The herders had their dogs, but were on foot for a half day while the camptenders were bringing the wagons and horses out from the stockyard in Wendover. While in Nevada, Will showed me the outside boundaries of the ranch, which had been shown to him by a previous owner.

Earlier, Mr. Dansie and I had agreed that keeping 7,000 replacement-ewe lambs would be about right for this year. We would keep the biggest and best. Our man in California could take only 2,000 that winter, give or take 30 head. There would be 5,000 to keep in Skull Valley. I would divide the 5,000 into three herds of about 1,700 each, and then put about 1,000 old blue-dot ewes in with each herd of ewe lambs. We could feed supplemental grain pellets to those three herds.

Mr. Dansie said there was a problem brewing, a disappointment. The new owners had now acquired enough company stock to be in control of major decisions. There would be no Christmas checks, so no employee would receive an extra one-twelfth of his annual earnings just before Christmas that year, 1952. I would have to tell this disturbing news to the men. And even worse, they would have to send a letter home before Christmas, with no check enclosed.

We shipped sheep to Skull Valley on November 12, 13, and 14. We sent three herds on the twelfth, three herds plus the California-bound ewe lambs on the thirteenth, and four herds on the fourteenth. The herders, brands, and counts for Skull Valley were:

November 12			
Dan	Black	●	2,874
Roque'	Black	1	2,900
Leandro	Black	+	2,951
			8,725
November 13			
Montoya	Red	2	2,886
Ross	Red	+	2,853
Velasques	Black	3	2,676
			8,415
November 14			
Calvillo	Red	0	2,537
Pacomio	Red	7	2,964
Adolfo	Red	T	2,746
Tony	Red	5	2,760
			11,007
Total to Skull Valley			28,147
Also Nov. 13, ewe lambs to California			2,029
Now in Nevada			11,424
Total ewes to be wintered			41,600
Rams			860
Total sheep to winter			42,460
Culls shipped to Omaha			1,714

In Nevada, Wally Denison had a pickup truck equipped to haul a saddle horse, so he could reach the far-distant areas of the range. He and I agreed on the days we would meet in Nevada to look over the situation. We found a place suitable for installing a 3,000-gallon water tank and 150 linear feet of watering troughs. This would be a big improvement to work on the following spring.

A few days later, Mr. Dansie told me that he had contacted Ferry Carpenter (Farrington R. Carpenter) of Hayden, Colorado, and asked him to sit down with us to discuss how best to handle the Dugway withdrawal problem. Mr. Carpenter was a knowledgeable range lawyer, as well as a cattleman. He organized the grazing districts in the West, and was the first director of the Bureau of Land Management. We could use his counsel. I met him in 1948, when he spoke at the National Woolgrowers' Convention in Salt Lake City. Mr. Carpenter was going to travel to California and had arranged to be in Salt Lake City at the Hotel Utah on December 4, 1952. Mr. Dansie and I conferred with him during the day. That evening, I was to have dinner alone with Mr. Carpenter at the hotel. The main conversation was how to document the cost of losing a sizeable chunk of the winter range, about 23 percent, to the Dugway Proving Ground. Those meetings helped.

A few days later, Mr. Dansie asked Marcellus Palmer, a private range consultant in Salt Lake City, to work with me on documenting the actual cost of this loss. We could work together, had known each other in college, and were both products of Utah State University in Logan. Palmer would be helpful. Mr. Carpenter again came to Salt Lake City a few days before Christmas, and stayed overnight at the Hotel Utah. That evening, he, Palmer, and I had dinner and a conference. We made some headway on strategy. We needed to study the situation some more, and decided to get together in a couple of weeks. On Saturday, January 3, 1953, Palmer and I left Salt Lake City very early and traveled to Craig, Colorado, to confer with Mr. Carpenter at his law office. We outlined information that we figured would be helpful in dealing with the federal appraisers.

After this meeting in Colorado, Mr. Palmer and I were to go to San Francisco to meet with representatives of the Army Corps of Engineers. Mr. Dansie suggested that we take our wives on this California trip. Mr. and Mrs. Palmer, Kathryn, and I drove to San Francisco in Palmer's automobile. Although we were on serious business, this was an enjoyable trip to a great city. The business meeting was friendly and constructive in some ways, but there was absolutely no give on the federal appraisal of $35,000. We Deseret Live Stock people came home empty-handed. It was agreed that there would be a meeting later in the spring, after both sides reconsidered the situation. I confided to Mr. Dansie that if the appraisal was changed, it meant someone had made a mistake. The government people were not about to let that happen.

The company was now deep into the process of changing ownership, so an outside "inspector" was hired to make an official count of all sheep and cattle as of February 1, 1953. He counted 41,874 sheep and 5,500 cattle. His sheep count was made when each herd was corralled and put single file through a

High-quality Hereford bulls, Skull Valley, 1953.

chute out on the range, in both Utah and Nevada, as we removed the rams at the end of the breeding season. As was usual at this time, a few thin ewes from each herd, along with the rams, were trucked to the ranch at Skull Valley. The rams were unloaded into their regular lot for feeding, and the thin ewes went into the scad herd for feeding on the range near the ranch. The usual work procedures and programs didn't change a great deal. We men butchered and processed beef and pork as always.

On February 21, I had a letter from Mr. Dansie saying that Marcellus Palmer and I should go to Colorado and meet with Mr. Carpenter on Saturday, February 28, 1953. At that time, we reviewed all the information pertinent to the damage sustained by the Deseret Live Stock Company in regard to Dugway. After we returned to Utah, calls were made to set a date for another negotiation in San Francisco. The earliest date that was satisfactory for everyone was over two months away, May 11 and 12, 1953. By then, the sheep would be shipped off the winter range and back to Wahsatch.

In early March, Art Krantz, Bill Gustin, and I installed a 3,000-gallon metal water tank and 150 linear feet of watering troughs at the selected site in Nevada. This strategic watering facility was north of U.S. Highway 40, and about fifteen miles south and west of ranch headquarters. This was a distinct asset to the Nevada operation. It could easily accommodate two winter herds, watering on an every-other-day basis.

President Henry D. Moyle conducted the annual stockholders' meeting on March 16, 1953. Mr. Dansie reported that the winter had been mild, and sheep losses were less than average. They reported the outside inspector's official count of both sheep and cattle as of February 1, 1953. The Deseret Live Stock Company's stock had largely changed hands, and now was the time to change officers and directors. Henry D. Moyle resigned as president, and James D. Moyle as vice-president. Walter Dansie would stay for a few months. Retiring directors were R. John Moss of Bountiful and William H. Sorensen of Spring City. Current directors James H. Moss of Bountiful and Edward O. Muir of Salt Lake City stayed on the new board.

The new owners were outstanding Utah businessmen. Kendall D. (Ken) Garff was elected president; David L. Freed, vice-president; and David A. Robinson, secretary-treasurer. New directors were Marriner Eccles, former chairman of the U.S. Federal Reserve, George Eccles, Marriner Browning, Harold Steele, B. H. Robinson, Simon B. Eggertson, and William Malecote. David Freed, in his remarks, complimented the retiring officers and directors for the great work they had done in building the company to its current size and financial condition.

Mr. Dansie had been general manager of the Deseret Live Stock Company since 1933. He had done a great job during those twenty years. The stockholders were told he would retire, probably in July, after shearing was completed. He would be succeeded by William Malecote. This arrangement would give the new manager an opportunity to work with Mr. Dansie and become well-acquainted with the company and how it operated. One day when I entered the Deseret Live Stock Company office in Salt Lake City, I stopped at the reception counter and said hello to Jim Circuit, the office manager, whose desk was just a few feet away. The door to Mr. Dansie's office was open. He and another fellow were seated at the desk, deep in conversation. Jim Circuit got up, came to the counter, shook hands, and said in a low voice, "That is Bill Malecote in there talking to Mr. Dansie." Jim told me that he was one of the new owners and was going to replace Mr. Dansie as general manager. Malecote was investing $220,000 in the company. He was in his early fifties, married, but never had any children. Jim thought they might be talking for quite awhile. I told Jim what I was trying to get done and said I would just go on about my business.

I did not meet Bill Malecote that day, but two days later he came to Skull Valley for a conference. As he approached, I could see that he was a bit under six feet tall, with only a slight bulge in the middle. He had a friendly smile, and a firm handshake. He said he saw me when I came into the office the other day and thought I would be around for a few minutes, but when Walt Dansie and he took a little break, I was gone. I told him that I left because Jim Circuit indicated that he and Mr. Dansie would be talking for a long time. Malecote asked if I thought they had made a mistake buying this outfit. I replied that from what I'd been told about the price, they'd made the buy of the century. This was a great outfit. I said it was just possibly the best sheep and cattle ranch in the world, and that the more he saw of it, the better he'd like it.

He told me that he would replace Walter Dansie as general manager, probably in July. They would be working together for now. He asked how often I went to Nevada, and I told him at least once a week, sometimes more often when something had to be done. It was a good piece of range. We visited about several things. Malecote was down to earth and said for me to call him Bill. He told me about his background—born in Tennessee, and went out West on his own when still just a kid. For four years, he was the bellhop in the Moore Hotel at Ontario, Oregon. By the time he was eighteen, he had a job at the local Ford agency, helping to get new cars into the showroom. He established good credit. Eventually Carl Graf, who owned the Moore Hotel, loaned him $5,000 to buy the Ford agency in Richfield, Utah. That was a good investment. The country was coming out of the Depression, and property values were going up. They sold out at a profit in Richfield and started buying ranch properties in Idaho, and later on in Montana. They would buy a place, fix it up, and sell it at a profit.

I said that I heard he was married, but had no children. He told me that he met his wife years ago, when he had surgery in a hospital and she was his nurse. They were both young, and she was good looking. She was a good partner, and a lot of his success was on account of her.

Then he said that he and the other new owners couldn't figure out why Walt Dansie and I got paid so much more than anybody else. I replied that Mr. Dansie was the general manager of one of the world's greatest ranches. He started out at $400 a month, during the bottom of the Depression in 1933. I didn't think $1,000 or $1,200 a month was too much for him. My salary was much less, but I'd been with the company seven years, and running forty thousand sheep was a big responsibility. I'd also been told to get familiar with the cattle operation both in Skull Valley and up at the Home Ranch. He cut me off and said they were not in the market for any bright young prospective manager. That was it for that day. We shook hands, and he left.

22

END OF AN ERA

THE APRIL SUN WAS BEATING down as I drove my Jeep pickup west across Skull Valley, heading for Fred Cordova's camp at the foot of Cedar Mountain, near Henry Spring, about four miles north of Eight-Mile Spring. I pulled up to camp, and Fred stepped out. I told Fred about the dates for shipping sheep off the winter range, and the herds scheduled for each day, so he would know who was just ahead of him, and who would be following. This was to prevent the herds from mixing. We visited a few minutes, shook hands, and I left. He was a top man.

In a few days, Mr. Dansie met Wally Denison and me in Nevada. We outlined plans for shipping two herds to Wahsatch each day, on May 1 and 2. The sheep would have to be trailed (driven) to the stockyards at Wendover, because loading required sturdy fences and corrals, with a loading ramp going to both the lower and upper decks of the double-deck railroad sheep cars. Ordinarily this would not be a problem, but the sheep had to cross busy U.S. Highway 40, and the stockyards were on the east end of town. Another problem was halogeton, a poisonous plant, growing on the range just west of Wendover. If the sheep were given a full feed of alfalfa hay before going through the section infested with halogeton, there would be no death losses due to eating this poisonous plant. Mr. Dansie told me that a couple of years previously, a sheepman whom he knew lost close to a thousand sheep due to halogeton. I was to make sure we had the two big trucks bring baled hay from Skull Valley, so our sheep were full before getting into the halogeton.

Will Sorensen went to Nevada to help with the shipping, and, as usual, Mr. Dansie counted the sheep into each railroad car. I went ahead to Wahsatch to receive the sheep. Shipping from Nevada went smoothly, and the counts were almost exactly what I had given to Mr. Dansie. May 3 was left open to get the crew and herds in Skull Valley ready for loading at Timpie on May 4, 5, 6, and 7. Will Sorensen arrived at Wahsatch as the last cars were being unloaded on May 8. I told Will that I'd been expecting Wally Denison every day since the sheep were shipped from Nevada, but he hadn't shown up. Will said that he wasn't coming. Wally did all right during the winter, but when the Nevada herds were loaded, Walt Dansie let him go.

All of the herds of ewes reached their designated lambing ranges before lambing started on May 10. Feed was plentiful—the grass came on fast. The numerous watering places that had been developed over the past several years were a big help. Although there were two days and nights of raw weather and freezing blizzards, the lambing was off to a good start.

I hated to leave for a few days right then, but I had to live up to the arrangements previously made regarding that meeting in California. Marcellus Palmer and I met with representatives of the Corps of Engineers on May 11 and 12, 1953, in San Francisco. This conference resulted in a lengthy discussion of the situation. We carefully reviewed all the costs incurred by the Deseret Live Stock Company in purchasing and operating the replacement property. However, there was no change in the federal government's offer of $35,000. The price had been set. A baseball umpire doesn't call "strike three," and then after a while say, "I'm sorry, it was a ball." An appraisal had been made, and the price set at $35,000. We men from the Deseret Live Stock Company were fighting a losing battle with the Corps of Engineers. To make a change in the $35,000 amount would indicate that the original appraisal was faulty, and this was not about to happen. The actual damage really was much higher, but it wasn't going to be paid for, except by the Deseret Live Stock Company. Palmer and I hated to report our failure to Mr. Dansie and the new owners.

It was a good lambing that spring of 1953, although cold storms took a heavy toll of newborn lambs. As docking time approached, our ten-year-old son Bill wanted to go with the docking crew and be a working member of it. He had been out several times in previous years and had helped catch the lambs and put them into the pen to be docked. This year he told me he wanted to be on the crew. He knew, and I knew, he could put the paint brand on the lambs after they had been docked. So I talked to Mr. Dansie, suggesting the company pay Bill a dollar a day for every day he worked during docking and shearing. During docking, we got up at five in the morning, and ate breakfast at five thirty. It was cold enough that everyone had to wear a coat. I told Bill to get to bed in good time each night, and I would wake him at five in the morning. I said he would be doing the work of a man. At five, all it took was a gentle tap on his shoulder and he was out of bed. It didn't take him long to get dressed and washed, and he was waiting for breakfast when the five thirty bell rang. As the crew filed into the dining room for breakfast, the high school boys and the other men all remarked that it was good to have Bill on the crew. He ate a substantial breakfast, and was the first man into the back of the truck.

That first day, I passed alongside him and said for him to check how many gallons of branding paint we had with us, and make sure we had the two small lamb brands we'd need that day. He replied that he had already checked. We had four gallons of red paint and brands for the two herds. He went docking all but one day, when he wasn't feeling well, and served as the branding man on the docking crew. Most days we docked more than 1,200 lambs in the forenoon, and about the same number in the afternoon. It was his job to put the paint brand on the back of each lamb. He was tired at the end of each day, but proud of the fact that he was doing a job that had always been done by a man or high school boy. He tended to business, didn't fool around, and everybody liked to have him on the job. Our crew docked over 25,000 lambs before shearing started.

At the beginning of docking, Sherman Hall, a high school boy from Manti in his second summer on the sheep operation, told me he would prefer working with cattle at the Home Ranch, where he could work all summer on the hay crew. This was arranged, and he transferred. In a couple of weeks, Pete Mower came over from the Home Ranch and told me that Sherman Hall had been killed the previous night, June 17, driving his own car between Evanston and the ranch. He had swerved to miss a cow out on the highway and rolled his car. Pete asked me to call Sherman's folks in Manti and let them know, and we visited a few minutes about the details. It was not an easy time for me. It would be tough on the boy's folks. I went to the house and told Kathryn. She had taught Sherman in a church class when he was a small boy. He was an active child, a "doer." He was a good, kind person, and a worker. I called Sherman's dad, Milton Hall, in Manti. Tears flowed. I met the parents later in the day at the mortuary in Evanston. It was a sad time. Losing a child tears at the innermost soul. My parents, Kathryn's parents, and the Hall families had always been good friends. This helped give a measure of heartfelt support. It would take the people at Deseret Live Stock Company a long time to get over this. The bereaved families never forgot. With shearing getting underway, I couldn't leave, and the Halls didn't expect me at the funeral.

Bill Malecote came for the start of shearing and planned to stay a few days. He observed young Bill at work and told me, "I like that boy. He works hard, and he tends to business." Bill Malecote told me that my son was a boy he would like to help someday. I replied that I was glad to hear that. Although now ten years old, Bill was only three when we arrived in the spring of 1946, I said. I'd taught him to work, and he'd grown up with the sheep.

The company men, not the shearers, were at the corral soon after daylight each morning, getting a herd of ewes and lambs partially separated before the six thirty breakfast bell. Usually when I came back to the house for breakfast, Dale and Doris were still sleeping. However Bill, now a dollar-a-day employee, and Diane were up navigating. Kathryn made sure each one had an enjoyable breakfast to start the day. Diane, now nine, and Doris, four, often went for a ride on Old Bally, but they came back so Dale, five, could ride out and help bring in the herds. Diane helped Kathryn, and was a constant companion for Doris.

Haler Witbeck came from California to serve as the south rider for the summer. He and Dale rode out together several times to help the herders bring their sheep to the shearing corral. One afternoon they had just corralled a large herd, and I was closing the gate. Haler walked up to me with a big smile and said, "You got a good helper in Dale." He told me that when they were still out there about a mile, Dale trotted up to him and said that he would take this herd on in if Haler wanted to go back for the next one. He really was good help, just like a man when he was on Old Bally and driving a herd of sheep.

Shearing was completed the first week in July. Mr. Dansie came to pay the shearing crew and write checks for the extra people who would not be employed during the summer. The last person to be paid was young Bill, who had worked thirty days and earned $30. This was a special occasion. Mr. Dansie told him, "This is the last check I will write for the Deseret Live Stock Company." As they shook hands, Bill said it was his first real paycheck. Mr. Dansie told me he was going to resign July 15, and would be going to Florida to manage the huge ranch operation being developed there under the name of Deseret Farms of Florida. He asked if I would care to go with him as a permanent employee of the Florida operation. I replied that my preference for the near future was to stay here as sheep foreman. I wasn't sure just how things would materialize with the new owners, but I really wanted to invest a little money in this outfit if I could and become a part of it.

Mr. Dansie then said, "Dean, I fear for your future in this company. I'm concerned for you and your family." He continued that under the present situation, the Deseret Live Stock Company was not my opportunity. When he hired me, he thought the company might well be a place where I could spend my life, and that working for it was a great opportunity for me to eventually replace him. The company had been good to him, and he thought it would be good for me. But the situation had changed, and the new owners would not want to pay me as much as I was currently making. I told Mr. Dansie he had been almost like a father to me. He and Will Sorensen had taught me how to manage. They made me grow. I was deeply indebted to Mr. Dansie for giving me responsibility and the opportunity to grapple with real problems. Few men had had the training that I had received. I said I was pleased to be asked if I wanted to go to Florida, but we had four children, and I wanted to raise them in the West.

Mr. Dansie said that I didn't have to decide right then about going to Florida. He was sure the Florida ranch would eventually run 25,000 mother cows, and more than that as the land was reclaimed and properly developed. He wanted Kathryn and me to know that he and Henry D. Moyle were pleased with us and our family The opportunity to help build the ranch in Florida was there, if we ever changed our minds. We talked a while longer, and I told him again how much I appreciated the training and guidance he had given me. He told me this was not the end of our relationship, as he was not going to stay in Florida permanently. When we got back to Salt Lake City, I was to give him a call at home. We would keep in touch.

With shearing completed, we shipped the cull sheep to Omaha, closed up the houses at the shearing corral, and migrated to the summer range. This was the eighth summer for our family at the Deseret Live Stock Company, but it was almost like entering a new world as we enjoyed the exuberance of being on top of the mountain in that luxuriant country south of Monte Cristo, at the head of Lost Creek. Will and Vera Sorensen moved their mobile home so he could supervise the work of Bill Watts and the fence crew. Tom and Jim Judd cooked for that crew and had a busy schedule preparing three substantial meals each day, every day.

My last letter from Mr. Dansie signed "General Manager" was July 12, 1953. My first letter from Bill Malecote signed "General Manager" was July 16. The transition of managers had been smoothly worked out over a period of several months. During the ownership and management transition of the Deseret Live Stock Company, the new owners hosted a dinner for new executives and key employees, and their wives, held at the roof-garden restaurant of the Hotel Utah in Salt Lake City. Among those invited were: president, Mr. and Mrs. Ken Garff; vice-president, Mr. and Mrs. David Freed; secretary-treasurer, Mr. and Mrs. David A. Robinson; new general manager, Mr. and Mrs. Bill Malecote; retiring general manager, Mr. and Mrs. Walter Dansie; foreman of the Home Ranch, Mr. and Mrs. Pete Mower; foreman of the Skull Valley Ranch, Mr. and Mrs. Glen Hess; and sheep foreman, Mr. and Mrs. Dean Frischknecht. Kathryn and I were proud to be associated with these comparatively young, successful businessmen who were the new officers and heavy stockholders.

It was an outstanding dinner and an enjoyable occasion. Ken Garff presided. First, he complimented the employees for the excellent condition of the sheep and cattle, and for the spirit of cooperation shown by everyone during the transition period. He then praised Mr. Dansie for his twenty years of dedicated and outstanding service to the Deseret Live Stock Company. Concluding these remarks, he reached into his pocket and pulled out two envelopes. He said that the company would like to present Mr. Dansie with a check, a going-away bonus of ten thousand dollars. Also, as a token of appreciation, they wished to present him with the title to the almost-new Cadillac he was driving. Although it was company-owned, it was now his to keep. These were generous parting gifts. Everyone was pleased to see Mr. Dansie receive the recognition and rewards. He had been an outstanding general manager, and a gentleman of good humor and keen judgment.

23

Bad News

I had invited Mr. and Mrs. Malecote to make a visit to the summer range while we were still at the shearing corral. In about a week they arrived, just after noon. They had already eaten lunch and could stay overnight. After a two-minute tour of the rustic facilities, they chose to sleep upstairs in the new, unfinished house. Bill Malecote asked me why we hadn't finished the house on the inside. I explained that last year Mr. Dansie told me to quit working on it. He said that the company was in the process of changing ownership, and he was told we should not put any more money into this house. It was basically a good house, but the stairway, inside walls on the main floor, and upstairs needed finishing.

The Malecotes were pleasant and easy to visit with. I wanted to show the new head man as much of the range as possible. That afternoon, we two men drove out in the pickup. First we went a mile north, through the log boundary fence, and looked over the range where two herds summered, mostly on national forest allotments. I explained that we had two herds out there, grazing for a little over two months, but the land was not included in the 250,000 acres considered to be the Home Ranch. We drove back through Scarecrow Gate, the metal gate in the log fence, then headed south on the main artery of a winding mountain road that covered the more than twenty miles to Cottonwood Flat, near the south boundary. We passed through beautiful groves of quaking aspen, grassy open vistas, and areas of shrubby browse country. We drove out a few miles on side roads: on the crest of Blue Ridge, Nelson, Monument, Horse Ridge, and Belknap, which drained east into Lost Creek, Hells Canyon, and the Weber River. We drove west on Lightning Ridge, Egan, Dog Pen, Knighton, Lake, and Bull Ridge, all gently sloping off toward Ogden.

I explained that after I came to work in 1946, it freed Will Sorensen to supervise Caterpillar bulldozers, improving these roads and building some two hundred watering holes on the summer range. Will had told me he would leave me better roads and more water than he ever had. Often Bill Malecote said, "Stop, let's get out and look." We gazed over miles and miles of country. It was almost unbelievable that these 250,000 acres were 90 percent private land. He was glad we had been building a fence around the outside, but he was going to cut all operating costs as much as possible. He would stop the fence crew right away, maybe in a week, and have Tom Judd and his wife move into a house at the Nevada ranch. Tom could do some irrigating and supervise the hay crew.

Kathryn prepared an excellent supper, and the Malecotes complimented her. Earlier she had told me she was concerned about trying to feed company, particularly new owners, in this crowded, one-room cabin. It held our double bed, single cots for Dale and Doris, the cookstove, and a table and chairs. We were crowded after the old two-room cabin had been torn down, and the new house wasn't yet usable. After supper, Malecote told me that he was going to have Will Sorensen come back here in early September and help get the lambs cut out and trucked to Ogden. We could figure out the dates, depending on the available feed. We enjoyed the Malecotes' visit.

When the fence crew was discontinued in July, Bill Watts moved his trailer to the north rider's camp, one mile south of headquarters, and resumed his responsibility as north rider. Tom Judd and his wife Jim had moved into the house at the Pilot Peak Ranch in Nevada, and both cooked for the hay crew. When haying was finished in August, Tom irrigated the hay fields, but he and Jim were not happy about living in such an isolated area. Jim did not drive an auto, and their nearest neighbors were fifteen miles away. Tom was sixty-two years old and knew he had better be looking for some other employment, as opportunities were limited.

On one of their trips north to Montello to pick up the mail, Jim was told that the Elko County, Nevada, school superintendent needed teachers for rural schools. She and Tom decided to investigate the situation. Tom reluctantly agreed to the possibility of Jim teaching school, and him keeping house and doing the cooking. They drove to Wendover, and Jim took the bus to Elko. She met the superintendent, and was offered her choice of five rural schools. She chose Ruby Valley and signed a teaching contract, beginning September 1, 1953. Tom then resigned from his employment with the Deseret Live Stock Company. Tom and Jim took a short vacation in Grantsville, purchased a new pickup truck in Salt Lake City, and moved to the house provided for the school teacher at Ruby Valley. They kept in touch with Kathryn and me.

On August 20, Kathryn and the four children moved home to Granger. Dale was five, too young for school, and wanted to be out on the ranch. I counseled our family that I could pick up Dale about September 10, and take him back to the mountain to help ship lambs, and then help make up winter herds at Salaratus. He could stay with me in my wagon until the middle of October. Everyone in the family agreed this would work, and I could handle it with the men.

Bill Malecote told me that during the last week of August, I should gather the rams out of their pastures and drive them to the shearing corral. He wanted to ship them to the company property at "E.T.," by Lake Point, where they could graze on grain stubble and residue in the fields for two or three weeks. We would have to give them a shot for "enterotoxemia," an overeating disease, before shipping. He hated to see that feed go to waste, especially in a dry year. I had an extra man help the buckherder gather the rams. When Dr. Osguthorpe, the veterinarian, came up from Salt Lake City to give the shots, he told me that we'd have to catch each ram and lay it on its side, so a rear flank was exposed. He showed me how to insert the needle between the layers of skin in the rear flank, and we two company men gave the rams the shots. This was exacting work. A misplaced shot into the gut or abdomen could cause death. We had over eight hundred rams, big and fat, with an average weight of 250 pounds. I had asked a nearby camptender, from six miles distant, to help us. There were five men working on this project, and it was a two-day job.

We waited a week for the medication to act before loading the rams onto the railroad at Wahsatch for their overnight ride to "E.T." After loading the rams, I drove my pickup, and pulled the herder's rubber-tired wagon to our home in Granger, so I could be with Kathryn and the children overnight. I left early, and was on hand to receive the rams the next morning. When the rams, the herder, and his camp were relocated in the designated area, I dropped in at the company office in Salt Lake City for a conference. Bill Malecote said we'd go get a sandwich. At a nearby café, a small newsboy about ten years old stopped at my side and asked if I wanted to buy a paper. I told him no, because we took that paper at home. Malecote asked the boy how many papers he still had to sell, and the price. He had three, at ten cents apiece. Bill bought them, gave the boy fifty cents, and said "Keep the change." The boy smiled, said "Gee, thanks Mister," and went on his way. Malecote told me that he knew how hard it was for kids to earn a little money. If they were willing to work, he was ready to give them encouragement. He gave that little guy a lift and taught me a lesson.

We discussed prospective dates in September for shipping lambs to market in Ogden. Malecote said that he would line up the trucks. I would be back in Salt Lake City before then, so we could make final arrangements. He would keep an eye on the rams. When the feed was gone, he would have the buck-herder drive them to Skull Valley. Malecote and I were very concerned about the shortage of feed on the winter range. During the summer, western Utah had experienced a severe drought, and now was declared a federal disaster area. This area included all of the Utah winter range in Skull Valley and on both sides of Cedar Mountain. Still vivid in my mind was that disastrous winter five years ago, when feed was short and the snow too deep. We didn't want a repeat of that situation. Malecote told me to keep only about 3,000 replacement-ewe lambs that fall. When I was cutting out the sale lambs to be trucked to Ogden, I'd keep about 3,000 of the biggest ewe lambs. He only wanted about 36,000 sheep wintered. That would make winter herds of about 2,200 sheep each, not 2,700 to 2,900 animals in each winter herd like we had been doing.

Will and Vera moved their mobile home to summer headquarters. He helped me cut several truckloads of firewood, and Art Krantz hauled it to the shearing corral. Before lamb shipping started in the middle of September, I hauled a pickup load of pine fireplace wood to our home, and had a good over-night visit with Kathryn and the kids. Dale returned with me to the summer range. Vera cooked for Will, Dale, and myself in the cabin where Kathryn and I had lived. She fixed huge breakfasts, with various combinations of hotcakes, eggs, ham, bacon, potatoes, cereal, and fruit. We ate at 5:00 a.m. and were at the corrals before daylight. Usually Dale got up early for breakfast and went with Will and me all day. Occasionally he said that he would stay and help Vera that day. The truckers left their homes "early-early," and were at the cor-rals ready to load lambs at daylight. Things went well, and all the lambs were shipped by the end of September.

One day, while still on the high country, Dale stayed to help Vera. Will and I had been out south in the pickup and were returning to headquarters. When we were only a quarter mile from home, all at once the six extra saddle horses exploded from the aspens, charged across the dusty road about a hun-dred yards in front of the truck, and galloped at full speed across the clear-ing to Lost Creek. Will said, "Stop! There's something after those horses!" I stopped, and was reaching for the .30–.30 in the gun rack, when out of the trees came Dale on Old Bally. He had a tight rein on the fired-up old horse. Old Bally wanted to run, and was dancing sideways. Dale was holding him down, sitting snug in the little saddle, his hat pulled down tight, while the old horse was still trying to get his head. "Dale," I yelled, "wait a minute." As the Jeep pulled up, Old Bally quieted down. I asked what was going on, and Dale explained that he was bringing the horses in to give them some oats. I told Dale that that was a good idea.

Will and Vera planned to terminate his employment at the end of the summer. They left in late September and returned to their home in Spring City. Will had been with the company for many years, and he really wanted to fully retire. He had done a great job. When we moved off the summer range in early October, the next scene of action was at the corrals at the dipping vat in Salaratus. Emma Zabriskie came to cook for us, and she served well-prepared meals. We were happy.

The last of October, when the market sheep were sold and gone, Bill Malecote called me aside for a private conference. He told me that he had voluntarily taken a $200-per-month pay cut from what Walter Dansie was being paid, and some of the new owners thought I should take a similar cut in pay. I was shocked. I said that I didn't feel that I could absorb a $200-a-month cut. Right then, my decision was that I'd stay at my current salary for a month, or six months, or a year, or indefinitely. I didn't want to leave the Deseret Live Stock Company and have to say to prospective employers that my salary had been cut by $200 a month. I suggested that we get the fall work done and the sheep shipped to the winter range. We could both be thinking things over, and then come to the best decision.

He agreed to go ahead with the operation of the outfit as usual. We shook hands, and each managed a smile. Then Bill Malecote drove back to Salt Lake City. I helped the men push the two extra covered wagons into the warehouse, and we drove back to the shearing corral. Alone now, washing up for supper in my wagon, I thought about my situation. I had a wife and four kids. I was thirty-three years old. Financially, we were okay for a while. The basement in our home had a year's supply of food. We could stand an emergency for a few months. In the meantime, I had my hands full running this outfit.

LIFE'S BIG DECISIONS

BECAUSE OF THE FEED SHORTAGE on the winter range, shipping was delayed by two weeks for the Nevada herds and even longer for those going to Skull Valley. We had kept only 3,200 replacement-ewe lambs, and our total inventory was now 37,000. The newly hired sub-foreman going to Nevada was "Big Joe" Guerricabeattia, six feet four inches tall, a dark, handsome Basque from Spain who had worked several years for Hatch Brothers, a close neighbor on the Utah summer and winter ranges. He had recently quit the Hatch outfit, and was "looking." However, I talked to Alvin Hatch before offering Big Joe employment. I told Alvin that I thought Big Joe wanted to go back and work for Hatch Brothers. If they wanted him back, I didn't want to interfere. Alvin replied that if Big Joe wanted to come back and work for them, he knew what he had to do. If we could use him in Nevada, Alvin said to go ahead. Big Joe had had plenty of time to come back.

On November 14, we loaded four herds going to Nevada, and the next day sent one more herd to Nevada and three to Skull Valley. There was still abundant feed, and only a little snow on the fall range adjoining Wahsatch. Some herds were held there as long as possible, four weeks longer than usual. I told Bill Malecote that we had culled far deeper than usual, and the herds were just beautiful. Anyone who knew sheep could tell at a glance that each herd was made up of excellent ewes. We had fat sheep. If we managed properly, we could winter them in good shape, in spite of the poor feed situation. Two herds, of almost 2,500 sheep each, would winter at the Home Ranch. Malecote said that they looked good. The quickest way he knew to improve a herd of sheep or cattle was to cull off the bottom end.

Malecote had purchased 120 young Hampshire rams. He told me that he wanted me to put those blackfaced rams in earlier than usual with our two herds of old blue-dot ewes at Eight-Mile. He wanted to try something that hadn't been done here for a long time, putting them in for ten days, November 15–24, and then pulling them back out. He wanted to try lambing these two herds in April here in Skull Valley, so I had to pull the rams back out. Otherwise, we would be getting lambs around shipping time. The young rams and aged ewes were in strong shape and engaged in an active, ten-day breeding season. Then the herds were corralled, with the rams being separated out and returned to the ranch. They would go out again December 15 and 16 for the regular breeding season.

Kathryn and I were much concerned about our future at the Deseret Live Stock Company. My annual salary of $7,200, plus what used to be an extra $600 at Christmas, for a total of $7,800, was about to become $4,800 if we agreed to stay and accept the proposal put forth by Bill Malecote. I needed to look for a more promising career. The situation wasn't just disturbing, it was alarming. I could not take time off to look around. I was well paid and still wanted to do more than was required. Opportunity had always come to me.

One day at the company office, Bill Malecote said that he and I should walk over to First Security Bank. He wanted to introduce me to Harold Steele, the head man there at the bank, and set up a personal checking account for me. He wanted to help me if he could, whether or nor I left the company. He said that he and I might still put our names together on a piece of paper some day.

The Dugway settlement was still hanging in the air. The company's people were understandably quite anxious to resolve this controversy over the government's proposed offer of $35,000 for our losing land, water, and grazing permits to the Dugway Proving Ground. Company executives asked Utah's U.S. senator, Arthur V. Watkins, to look into the situation. It was pointed out that the replacement property in Nevada had cost $215,000. The company figured $121,360 would be fair compensation for what was actually lost at Dugway. The case was reviewed, nothing happened, and time passed. On December 7, 1953, the Deseret Live Stock Company proposed the following settlement amount:

Loss of sheep grazing	$49,071.11
Loss of cattle grazing	5,100.00
Loss of springs	10,000.00
Loss of improvements (corrals)	500.00
	$64,671.11

A short time later, the issue was finally settled when the Deseret Live Stock Company agreed to accept a total of $41,000 in complete settlement for losses caused by the "Dugway withdrawal."

One day Bill Malecote came to Skull Valley and met me at the ranch in the late afternoon. He had several things on his mind. He had talked to Will Sorensen about me. Will said that there was no question that I could run the sheep better than any man alive, but that I had a wife and four kids that were going to demand more of my time. Will thought the Deseret Live Stock Company might be better off with a sheep foreman who didn't have a family of growing kids, someone who could live on the outfit twenty-four hours a day, like he did. I was surprised to hear that about my family. I'd given full time to this outfit and knew I could run those sheep about right. I'd been through nine shearings, and this was my eighth winter. By now, I ought to have been able to run them. I added that the company had quite a financial investment in me, and I had invested nearly eight good years in the company.

Malacote said the other thing he wanted to know was if I was willing to take a pay cut to $400 a month. I told him that I realized prices weren't as good as they had been, and that I didn't want to leave the Deseret Live Stock Company. On an outfit of this size, the $2,400 difference in my salary would not matter a particle to the company, but it made a tremendous difference in the quality of life for my family. I didn't think a $7,200 yearly salary was too high for my responsibilities. He said that he would tell the board of directors what my decision was. He might have to replace me with someone who would accept less money, and said he could get an experienced sheep man for $400 a month. Malecote asked whether, if I left the company, and he needed help, I would come and help him. I replied that I would, but that I didn't want to leave the company. We shook hands firmly, and each of us managed a smile.

Hog butchering was accomplished, as usual, in early December, and then the rams were put into the herds. Things were moving well. A few days before Christmas, I returned from Nevada to the ranch in Skull Valley one afternoon to find a new man there, Bill White, fresh from Colorado. As we walked over to the big house for supper, he explained to me that he came from Colorado to the company office in Salt Lake City, and then rode here to the ranch with Art Krantz, who'd been to town for a load of equipment and supplies. Bill White was a pleasant young man, about my age. He had a wife and family in Colorado. We had no need of extra men at the time, so I figured White was my replacement. White said that he was hired by Bill Malecote. He was to ride with me over the winter ranges in Utah and Nevada, and learn all he could from me about how the total sheep outfit was managed. I took him in the pickup each day and endeavored to show him everything. One man's career at the company was expiring, and one man's just beginning. I knew Bill White was going to be taking over a tremendous responsibility, without the close guidance that had been given to me by Will Sorensen and Mr. Dansie.

From left, Doris Frischknecht, Gene and Kay Gelb, and Diane, Dale, and
"Bill" Dean Frischknecht at the Deseret Live Stock Company's south
ranch in Skull Valley, November 1955.

I felt sorry for this man, entering an unfamiliar, yet demanding, situation. I told him that if he needed help after I was gone, to call on me. I'd told Bill Malecote I'd help if needed.

At home for the night in Granger, Kathryn and I spent considerable time discussing our situation. We had helped the company, strengthened the weak spots, and assisted it to build and expand. My major responsibility, the sheep operation, was much more productive than when we first came here. Kathryn had helped me make a worthwhile contribution to the company. The next night, alone in my camp at Skull Valley, I wrote the following to myself: "Sadly, to me, my career in this company is coming to an end, not the end I anticipated, but still, an end of my choosing. I want to leave here as a vital part of Deseret Live Stock Company at its peak. Part of me stays here."

We counted all the herds for an official year-end inventory. As of January 1, 1954, the company had 36,694 ewes. Cattle numbers had slightly increased, to 6,250. It was evident that in the near future Bill Malecote would notify us about what was going to take place. The first week in January, he called me in Skull Valley and gave me a date to arrange a four-way conference "for three Bills and one Dean." It would be Bill Malecote, Bill White, Bill Watts, and me. When we met in Skull Valley, Malecote told us that as of that moment, Bill White

From left, Doris and Diane Frischknecht, Kay Gelb, Dale Frischknecht, Gene Gelb, and "Bill" Dean Frischknecht; at the new main house at the Deseret Live Stock Company's ranch at Iosepa, Skull Valley, November 1955.

was replacing me as sheep foreman for the Deseret Live Stock Company, and I was to turn the records over to him. He already had a list of all the herders and camptenders, and of how many sheep were in each herd. I was to go to the company office in the next day or two and write a letter of resignation. The company would pay me through January. I told them that I would come back and help if needed. Bill Malecote then went back to Salt Lake City.

I gathered up my belongings from the almost-new camp I had been using for the last two years and put them into my personal auto. I shook hands with Bill White and wished him well. With Bill Watts, it took longer. We had worked together for almost eight years, and had shared some rough, tough days. We had also shared the good days. We could depend on each other to do a job, and had a mutual, deep respect. Running more than forty thousand sheep, day in and day out, took a lot of dedication, as well as "know how."

On January 8, 1954, I went to the company office in Salt Lake City and wrote a letter of resignation. Those were significant days—I was changing my career. It was a hard thing for me to leave the Deseret Live Stock Company. I felt as if my world was coming apart, but not for long. Now was a time to assess the big picture, count blessings, and move ahead. Bill Malecote was not in the office. However, I wrote a nice, but brief, letter of resignation addressed to him as general manager. It said: "After carefully considering the present situation at Deseret Live Stock Company, it is my decision to resign as Sheep Foreman. My years with the Company have been enjoyable. We have made many improvements on the range, and in the general operation of the sheep. The sheep are in excellent shape. Best wishes to all of you at Deseret Live Stock Company." I then signed my name to it. The office manager, Jim Circuit, typed this letter and said he would give it to Mr. Malecote.

Bill White had confided to me that he was not a completely well man, and had some physical problems. This situation, coupled with the sudden responsibility of running the company's enormous sheep operation, must have been a terrible strain on him. He had been on the job for only a few weeks when he failed to show up for breakfast at the ranch one morning. He was found dead in his wagon, the victim of natural causes. I learned about this a couple of days after it happened. I liked Bill White and was sorry to learn that he had died. Bill Malecote called me one evening soon after this happened. He said that they had a problem in Nevada with regard to a boundary line. He asked me to go out and confer with the company's men in Nevada, as well as with the offending neighbor, and get the situation straightened out. I told him that I would be glad to do so. He said for me to report the situation to Bill Watts, who was at the ranch in Skull Valley, on my return, and then come in to the company office so they could give me a check for my time and travel expenses.

I made the trip to Nevada, resolved the misunderstanding, and drove up Skull Valley to the ranch. Bill Watts was there, so I explained what the problem had been in Nevada and said it was taken care of. Bill Malecote was glad for my help and paid me for time and travel. I then said, "If you're in a bind for foreman, I'll come back and help." He said, "I'll take you back." However, we needed to walk over a few blocks and run this by another of the new owners. The other man said they had agonized over the decision about me, but had decided they couldn't afford my salary. He thought they would have to live with that decision. We parted friends, but inwardly I was having a hard time controlling true sadness about severing the knot. Bill Malecote said, "Come see me occasionally. We still may sign a paper together some day."

On March 8, 1954, Will Sorensen came out of retirement at Spring City, and returned to Skull Valley on a temporary basis as sheep foreman for the Deseret Live Stock Company. Later that spring, he was replaced by Ted Bennion.

TRANSITIONS

AFTER LEAVING THE DESERET LIVE Stock Company I lost no time in contacting Ivan Johnson, manager of his own insurance agency for Pacific National Life in Salt Lake City. Some time previously, he had suggested that if I ever left the Deseret Live Stock Company, he would be pleased to have me go to work for him. Mr. Johnson conducted a training session for me and other new agents. I completed the first year of the LUTC (Life Underwriters Training Course), was making a living, and learning a lot.

Mr. Johnson gave me a quota to sell by June 1, 1954. If I accomplished this, he would pay all expenses for Kathryn and me to attend the 1954 annual Pacific National Life Convention at the Broadmoor Hotel in Colorado Springs. I sold the quota.

It was announced that the 1955 convention would be at the Del Coronado Hotel in San Diego, and the 1956 convention would be in Honolulu, Hawaii. All insurance sold during the next two years would go toward qualifying for both conventions. Mr. Johnson gave me a quota that would qualify for the 1955 Pacific National Life Convention in San Diego, with all expenses paid for Kathryn and me and our four children to attend the convention. I qualified, and it was wonderful. I also qualified for the 1956 convention in Hawaii, with all expenses paid for Kathryn and me to attend that convention at Waikiki Beach, Honolulu.

In the spring of 1955, I chanced to meet Bill Cook on the street in Salt Lake City. We visited and brought each other up to date. He was buying wool for G. A. Hanson, owner of Wool Handlers, Inc., in Salt Lake City. We went to see Mr. Hanson about hiring me to buy wool on a part-time basis, during the spring shearing season. I had known Mr. Hanson for several years. He had bought wool and lambs from my parents at different times. He told me what my commission would be, which was good. Then he got me properly licensed with the state as a "wool broker." I bought wool again for Mr. Hanson in early spring, 1956.

Previously, back in September 1955, Frank Ballard, director of the Oregon State College Extension Service, had interviewed me in Salt Lake City about becoming an extension livestock specialist. Although that position was vacant, they were not going to fill it for a few months. In April 1956, Mr. Ballard called me, and asked if Kathryn and I would have lunch in Pocatello, Idaho, with Mr. Jean Scheel, the assistant director of the extension service. It was a pleasant occasion. Near the end of the lunch, he said that Oregon State was ready to hire me. When I told him of the forthcoming trip to Hawaii in June, he was pleased. He said for me to figure on coming the first of July. My career change went smoothly and efficiently. I was appointed an extension livestock specialist, headquartered at what is now Oregon State University in Corvallis, Oregon, but with responsibilities throughout the state. Advancements were frequent and generous. We had made a wise decision.

Kathryn and I and our four children returned to Utah each summer for family reunions and visiting. Occasionally, we ate lunch with Mr. Dansie in Salt Lake City. He stayed at our home in Corvallis when he was in the area on church welfare business. We kept in close touch with Will and Vera Sorensen at their home in Spring City. We always reminisced about the old days and were updated on the Deseret Live Stock Company. My time working for the company was a most valuable experience.

Bill Malecote did not stay long as general manager of the company, less than three years. The main owners, representing their families' interests, were: Ken Garff, about 35 percent; David Freed, about 35 percent; and David Robinson, about 20 percent. I had been gone from the Deseret Live Stock Company for twenty years when these owners sold the major portion of the company to Joseph Hotung in 1974. The sellers retained ownership of Heiner's Canyon, 26,000 acres of private range on the southeast part of the Home Ranch, and 2,000 mother cows, which summered in Heiner's Canyon and were trucked to Skull Valley for winter grazing. Garff, Freed, and Robinson kept the two ranches in Skull Valley, and converted much of the BLM sheep winter grazing permit in Skull Valley to a cattle permit, to accommodate the extra cattle.

Joseph Hotung was from a wealthy family in Hong Kong and had citizenship both there and in the United States. He purchased the Home Ranch, 226,000 acres bordering the Wyoming state line, which included the high summer range. The ranch now had 201,000 acres of private land, and about 25,000 acres of federal land. Hotung's purchase included two thousand cattle and ten thousand sheep. The cattle stayed on the Home Ranch year around.

The sheep summered in the high country of the Home Ranch and wintered on their usual range west of Cedar Mountain.

While on a visit to Utah in 1983, I called the Home Ranch of the Deseret Live Stock Company to talk to the general manager, Greg Simonds. He told me that the last of the sheep were sold during the Hotung years. The LDS Church had recently purchased the ranch from Joseph Hotung, and the name was now the Deseret Land and Livestock Company. Greg invited me to dinner at noon the next day at Home Ranch headquarters. The dinner was being held in honor of the seventy-fifth birthday of Pete Mower, former manager of the Home Ranch, now retired. We had a most enjoyable visit, discussing times long gone. Later, I rode with Greg in his pickup over some of the lower range, including a visit to the old sheep-dipping-vat corrals at the mouth of Black Dan Canyon. The unused corrals were now just a few rotting logs. The small cookhouse we built back in 1947 was a home on the range for the cowboy riding in that area. The house built of sawn logs, with a shingled roof and tongue-and-groove board flooring, was holding up as though it would last a long time.

In 1985, Kathryn and I purchased my parent's home in Manti, Utah. This belonged to my mother's parents. We spend considerable time there. This helped me keep in touch with previous owners and employees of the Deseret Live Stock Company. Past-president Ken Garff, past-vice president David Freed, and past-secretary David Robinson helped to keep me informed.

In the spring of 1991, I was in my office in the Animal Science Department at Oregon State University, when I got a telephone call from Greg Simonds, speaking from his office at the Home Ranch in Utah. He invited me and our family to help celebrate the hundred-year birthday of the Deseret Live Stock Company, to be held on August 24, 1991, at Home Ranch headquarters. I thanked Greg and told him we would be there. It had been more than thirty-seven years since I worked there. I had often reflected on those old days—the good, the bad, the humorous, and the tragic. We at the old Deseret Live Stock Company had lived a happy but hard, rugged life, with hopes, ambitions, fears, and desires that drove us to try again and again. I had kept accurate records, and had written about some of what happened.

On the day of the celebration, when Kathryn and I and two of my younger brothers and their wives, Kay and Jean from Manti, and Clair—who had worked on our fence crew the summer of 1947—and his wife Jean from Ogden, all arrived at the old Home Ranch of the Deseret Live Stock Company, we mingled with a festive crowd of over five hundred people. These were previous stockholders, former and current employees, and associated family members.

It was a great occasion, honoring the first hundred years of one of the world's greatest ranches. Greg Simonds and his wife Julie welcomed us, and we toured through the main floor of the big house. I had a good visit with Ken Garff, who was president of the company when I left in 1954. He told me that David Robinson had died two years previously. That left the Garffs and the Freeds as owners of Heiner's Canyon and the property in Skull Valley, as well as the cattle they kept when the big sale was made to Joe Hotung. Recently, Ken had sold the Garff family interest to the Freed family. Ken went on to say he was eighty-five years old, exercised an hour each day, and was in excellent physical health. He went to work each weekday at his office on State Street and Sixth South in Salt Lake City. He said, "Dean, come see me when you can."

Greg Simonds and his crew had the day well organized. He told the crowd that five buses were present to transport guests on two tours, one to the old shearing corral over close to what used to be the small railroad town of Wahsatch, and the other to the lower part of the summer range, going as far as the roads would allow bus travel. The buses stopped close to a herd of five hundred mature buffalo, out south about three miles. People had to stay on the bus, but we marveled at the fearless buffalo as they ambled across the dirt road between buses parked fifty yards apart. Greg Simonds, on our bus, explained that the buffalo grazed on the range year-round. A little supplemental hay or pelleted feed was given to the buffalo during part of the winter. He tried to manage with adequate feed so the buffalo cows could produce a 90 percent calf crop. He went on to say that there was a strong demand for buffalo bull calves, sold as potential breeding stock. Surplus yearling and two-year-old bulls could be sold for meat for human consumption. A few mature old bulls were allowed to be harvested by hunters, shooting them out on the range.

At the shearing corral, the old houses we had lived in were gone. They had been bulldozed into piles and burned long ago. Where the old log cookhouse had formerly stood, there was now a substantial building, a hunting lodge. It was the sturdy, well-constructed, Union Pacific station building, moved the three miles there from the depot at Wahsatch. The railroad hamlet, Wahsatch, no longer existed. It was one of the stations closed up and disposed of as the Union Pacific streamlined its operations. I told our tour group that I wished I could stand by the station again as Bob, the station agent, stood by the side of the track. He would have a message fastened to a looped string on the end of a long stick, which he hoisted into the air. The engineer, coming at fifty miles an hour, would stick his arm through the looped string, and take the string and message on board the train.

Back at Home Ranch headquarters, we enjoyed a sumptuous steak dinner and some excellent entertainment. It was a great day of visiting and

Author's Collection
The enlarged new home at the former sheep summer headquarters, now a hunting lodge. From left, Anne Merten, Paul Merten, Greg Merten, Diane Frischknecht Merten, and Dean Frischknecht, 1994.

remembering days long ago. However, our old-time, vast, forty-thousand-sheep operation is no more. The company owns no sheep, although a small part of the range is leased to other sheep producers. From sheep to big game is a major transition. The ranch is also home to 4,500 mother cows.

On August 3, 1994, when several of our Oregon family were in Utah, I accompanied my daughter Diane, her husband Greg Merten, and their two youngest children, Paul and Anne, on a trip to the Home Ranch of the Deseret Land and Livestock Company. Jay Olsen, from Manti, was a range scientist at the ranch. He escorted us to the high summer range and to our former sheep headquarters. The new home, which we did not get finished before I left the company in 1954, had been enlarged and completed, with bedrooms on both floors. On the south side of the house, a substantial new porch, built of heavy logs, added strength and rustic charm to the front entrance. Nowadays, no one lives all summer at this location, because the sheep herds are no more. The house serves as a hunting lodge and has inside plumbing. Just a few rotting logs remain of the once-proud fence. Times change, and that log fence is no longer needed. The low wire fence now in place allows easy jumping for the thousands of deer and elk grazing this area.

This was a time of nostalgia for Diane and me. We recalled our horse rides and our trips in the pickup over these dusty roads. The plentiful summer dust hadn't changed. As we were leaving our long-ago, high country headquarters, Diane said that it was good to once again see the distant horizons of where she and the others grew up. Our family was isolated here during those golden summers in these far out spaces. The far horizons helped us to see the big picture of life more clearly.

Descending the mountain back to the Home Ranch, we took the north road down Five Spring Canyon. Passing through a grove of quaking aspens, we surprised a huge bull moose, drinking at a pond a few yards off the road. He raised his head and looked us over, but wasn't impressed enough to be stampeded. Moose numbers are increasing on this ranch. Back down at the ranch, we were invited to tour through the big house, which is always a pleasure. I have always enjoyed standing on the east porch and gazing across the miles of native-grass meadows. The well-kept barns, fences, houses, and yards are a credit to the people now running this outfit. People live, work, and die. The land resource remains.

About a year later, on July 31, 1995, my nephew Paul Frischknecht of Manti, then president of the Utah Woolgrowers' Association, invited me and my granddaughter, Anne Merten, to accompany him and his daughters, Julie and Rachelle, on a two-night trip to the Deseret Land and Livestock Home Ranch. Close friends Anne and Julie were thirteen, and Rachelle was younger. The occasion was a two-day board of directors' meeting of the Utah Woolgrowers. We were comfortably quartered on the main floor of the big house. The girls had a room with bunk beds, close to a room with two beds that Paul and I shared. Several of the directors and their families occupied rooms upstairs. Early the next morning, Paul and the girls joined a group going to the high country in pickup trucks for a two-hour tour. They returned on time, and the girls enthusiastically reported that they saw twelve elk up close, and a large herd in the distance, too many to count. They saw six deer up close, a mother cow moose with twin calves, and a yearling moose.

By now Greg Simonds had been elevated to a supervisory capacity over this ranch as well as three or four others in Wyoming and Montana. Bill Hopkins was now the resident manager of the Home Ranch. Rick Danver had primary responsibility for the wildlife management. Jay Olsen was a tour guide for part of each day, when meetings were not in session. He told us the herd of buffalo had been sold. Low wire fences—which elk, deer, and antelope can jump—turn and control the cattle. The buffalo just walked through the fences and created too many problems. While we were at the ranch, we were fortunate each day to have three meals served in the dining room of the big house. The Hopkins and Olsen families, along with Rick Danver's wife Sylvia, and Missie Argyle from nearby Evanston, Wyoming, prepared our meals, and they did a professional job.

Meanwhile, what about the 360,000-acre Deseret ranch southeast of Orlando, Florida? During the summer of 1984, Kathryn and I attended the annual meeting of the (international) Beef Improvement Federation (BIF) in Atlanta, Georgia. I was particularly interested in seeing this ranch, since back in 1953, I had chosen not to go to Florida when Mr. Dansie offered me the opportunity to help develop the ranch. Following the BIF meeting, we drove south to the Deseret ranch, and spent most of a day with general manager Paul Genho. Paul had this outfit divided into twelve individual ranches, each with its own crew and facilities for running 2,500 mother cows, for a total of 30,000 cows for the whole operation. I have kept in touch with the place in Florida. The ranch is now home to 40,000 mother cows, the largest number of cows on a single ranch in the United States.

What now remains as the Deseret Land and Livestock Company in northeastern Utah is possibly the finest example in the world of a great ranch, combining land, livestock, and wildlife. It has had many honors and recognitions for top management of these resources.

Staying in the West, and eventually working for Oregon State University, had worked out well for our family. When our Oregon great-grandchildren stay at our home in Manti, Utah, each summer, they are the sixth generation of our family to occupy this sturdy, cut-stone home. I worked for nearly eight years at the Deseret Live Stock Company, 1946–54. That experience was a great influence on my life and on the lives of Kathryn and our children. It was a building and strengthening experience that helped each of us. Those times are impossible to forget. And it is a ranch to remember.

INDEX